National
4 & 5

Geography

Global Issues

Calvin Clarke
Susie Clarke

HODDER
GIBSON
AN HACHETTE UK COMPANY

The Publishers would like to thank the following for permission to reproduce copyright material:

Photo credits

Chapter opener image on pp.1, 7, 13, 19, 22, 27 and 33 © tusharkoley – Fotolia; p.5 (top left) © S. Wanke/Photodisc/Getty Images, (top right) © Arctic-Images/Corbis, (bottom left) © tusharkoley – Fotolia; p.6 (left to right) © Siede Preis/Photodisc/Getty Images, © Siede Preis/Photodisc/Getty Images, © mirpic – Fotolia; p.9 NASA; p.15 © V. ZHURAVLEV – Fotolia; p.17 (top left) © Mikael Damkier / Fotolia.com, (top right) © Elwynn - Fotolia.com, (centre left) © Elwynn - Fotolia.com, (centre right) © Lubos Chlubny – Fotolia, (bottom left) © david hughes – Fotolia, (bottom right) © Sergiy Serdyuk – Fotolia; p.18 (top left) © eag1e – Fotolia, (top right) © TMAX – Fotolia, (bottom left) © corepics - Fotolia.com; p.28 © jjayo - Fotolia.com; p.34 © Getty Images/Stockbyte Platinum/Thinkstock; p.37 © Pat Canova / Alamy; Chapter opener image on pp.39, 42, 47, 53, 59, 62, 69, 74, 80, 85 and 90 NASA; p.55 U.S. Geological Survey; p.57 © Corbis. All Rights Reserved.; p.60 (top left) U.S. Geological Survey/photo by D. Dzurisin. Cowlitz County, Washington. May 31, 1980, (top right) © Bart Rayniak/The Spokesman-Review, (bottom right) © David R. Frazier Photolibrary, Inc. / Alamy; p.61 © Gary Braasch/CORBIS; p.75 (top) © AlamyCelebrity / Alamy, (bottom) © Kyodo/Xinhua Press/Corbis; p.89 (top left) © Mark Wolfe/ FEMA, (top right) © Michael Rieger/FEMA, (bottom) © Marc Serota/Reuters/Corbis; p.91 © Jocelyn Augustino/FEMA; p.92 NOAA; Chapter opener image on pp.95, 101, 108, 112, 117, 121, 124 and 130; © Paul Cowan – Fotolia; p.109 eBay Mark is a trademark of eBay Inc.; p.114 © Mystic Arabia / Alamy; p.119 © PILAR OLIVARES/Reuters/Corbis; p.122 (left) © FOOD DRINK AND DIET/MARK SYKES / Alamy, (right) Mr Hicks46/http://creativecommons.org/licenses/by-sa/2.0/deed.en_GB/http://www.flickr.com/photos/teosaurio/4532254899/; p.124 © Crispin Hughes/Oxfam; p.127 (top) Reproduced with kind permission of the Fairtrade Foundation, (centre) © Rainforest Alliance, (bottom) © Paul Cowan – Fotolia; p.133 CEFICEFI/http://creativecommons.org/licenses/by/3.0/deed.en/http://commons.wikimedia.org/wiki/File:RENAULT_SAMSUNG_MOTORS.JPG; Chapter opener image on pp.136, 141, 146, 150, 154, 160, 165 and 168 © punyafamily – Fotolia; p.144 © Iamnee – Fotolia; p.145 (left) Image Courtesy of The Advertising Archives, (right) SunSmart Program/Cancer Council Victoria; p.147 © Henrik Larsson – Fotolia; p.152 © punyafamily – Fotolia.

Every effort has been made to trace all copyright holders, but if any have been inadvertently overlooked the Publishers will be pleased to make the necessary arrangements at the first opportunity.

Although every effort has been made to ensure that website addresses are correct at time of going to press, Hodder Gibson cannot be held responsible for the content of any website mentioned in this book. It is sometimes possible to find a relocated web page by typing in the address of the home page for a website in the URL window of your browser.

Hachette UK's policy is to use papers that are natural, renewable and recyclable products and made from wood grown in sustainable forests. The logging and manufacturing processes are expected to conform to the environmental regulations of the country of origin.

Orders: please contact Bookpoint Ltd, 130 Park Drive, Abingdon, Oxon OX14 4SE. Telephone: (44) 01235 827720. Fax: (44) 01235 400454. Lines are open 9.00–5.00, Monday to Saturday, with a 24-hour message answering service. Visit our website at www.hoddereducation.co.uk. Hodder Gibson can be contacted direct on: Tel: 0141 848 1609; Fax: 0141 889 6315; email: hoddergibson@hodder.co.uk.

© Calvin Clarke and Susan Clarke 2013

First published in 2013 by

Hodder Gibson, an imprint of Hodder Education,

An Hachette UK Company

2a Christie Street

Paisley PA1 1NB

Impression number	5	4	3	2	1	
Year	2017	2016	2015	2014	2013	

ISBN: 978 1 4441 8747 2

Cover photo © Michael Bocchieri/Getty Images

Illustrations by Emma Golley at Redmoor Design and Integra Software Services Pvt. Ltd., Pondicherry, India

Typeset in 11 on 12 pt Stempel Schneidler Std Light by Integra Software Services Pvt. Ltd., Pondicherry, India

Printed in Italy

A catalogue record for this title is available from the British Library

Contents

Introduction

This book has been written to cover Unit 3 of the Scottish Qualification Authority's National 4 (N4) and National 5 (N5) Geography courses.

Unit 3: Global Issues

In this Unit, learners will develop skills in the use of numerical and graphical information in the context of global issues. Learners will develop a detailed knowledge and understanding of significant global geographical issues.

Candidates should study **two** global issues from a prescribed list of six. Four are covered in this book.

Climate change

This topic is covered in chapters 1–7.

- Features of climate change
- Causes – physical and human
- Effects – local and global
- Management – strategies to minimise impact/effects

Environmental hazards

This topic is covered in chapters 8–18.

- Describe the main features of earthquakes, volcanoes and tropical storms
- Causes of each hazard
- Impact on the landscape and population of each hazard
- Management – methods of prediction and planning

Trade and globalisation

This topic is covered in chapters 19–26.

- Description of world trade patterns
- Cause of inequalities in trade
- Impact of world trade patterns on people and the environment
- Strategies to reduce inequalities – trade alliances, fair trade, sustainable practices

Health

This topic is covered in chapters 27–34.

- Describe the distribution of a range of world diseases.
- Explain the causes, effects and strategies adopted to manage:
  - HIV/AIDS in developed and developing countries
  - one disease prevalent in a developed country (choose from: heart disease, cancer or asthma)
  - one disease prevalent in a developing country (choose from: malaria, cholera, kwashiorkor or pneumonia).

Each chapter contains N4-level and N5-level questions and/or Activities. They are designed to develop knowledge and understanding of global issues but also to develop a range of skills. These include a range of research skills in Outcome 1 but also skills such as literacy, numeracy, enterprise, citizenship and thinking skills. The N5 questions are mostly similar in style to examination questions; the Activities test the same concepts but encourage active learning and the development of a wider range of skills.

Answers to questions at N4 and N5 are differentiated chiefly according to the amount of detail given. The Activities are differentiated by student response.

Each chapter in the book begins by stating the learning intentions. At the end of the chapter students are asked to self-assess their understanding of these learning intentions using the traffic light system. A photocopiable checklist for all the learning intentions is found at the back of this book for students to use. This 'I can do' self-assessment approach is explained to students on the following page.

'I can do'

Each chapter in this book has a box at the beginning outlining what you will be learning and what you should be able to do after you have completed the N4/N5 questions and/or Activities. For example:

It is very important that you feel confident about these as you will be assessed on them.

After you have completed each chapter, you will be asked to fill out the 'I can do' boxes for that chapter. These can be found at the back on pages 172–176.

The 'I can do' checklist outlines all of the learning intentions for every chapter within this book. You have to fill it out based on how well you understood the information in the chapter. It is a 'traffic light' system:

This chapter looks at the physical factors causing climate change.

By the end of this chapter you should be able to:

✓ describe how physical factors affect the world's climate
✓ give examples of some physical factors that cause climate change
✓ explain how these factors cause climate change.

Now complete the 'I can do' boxes for this chapter.

● **RED** means that you DO NOT FEEL THAT YOU UNDERSTAND THIS AND DON'T THINK YOU CAN DO THIS NOW.

● **YELLOW** means that you THINK YOU CAN DO MOST OF IT BUT YOU STILL HAVE SOME PROBLEMS.

● **GREEN** means that you FULLY UNDERSTAND THIS AND YOU CAN DO IT WITHOUT ANY DIFFICULTY.

On the 'I can do' checklist, there is also a space for comments. It is worthwhile taking a few minutes to write a few comments about the chapter, as they will prove very helpful when you start revising. See the example below.

	Red	Yellow	Green	Comment
Chapter 1 Climate change				
Describe what climate change is		✓		I understand what climate change is but I'm still unsure how to describe it concisely.
Describe the natural greenhouse effect			✓	I understand what the natural greenhouse effect is and can describe it accurately.
Give examples of evidence that our climate has changed			✓	I can give detailed examples of evidence that our climate has changed.

Climate change

This chapter looks at how the world's climate is changing.

By the end of this chapter, you should be able to:

✓ describe what climate change is
✓ describe the natural greenhouse effect
✓ give examples of evidence that our climate has changed.

Did you know...?
There are many different greenhouse gases, such as water vapour, carbon dioxide, methane, nitrous oxide and ozone.

The natural greenhouse effect

The Earth is surrounded by an invisible layer of air called the atmosphere. In the atmosphere there are many different gases, chiefly nitrogen and oxygen, but several others in much smaller amounts. They are all important to us in different ways. **Only some of the gases in the atmosphere are greenhouse gases**; but they are essential because they help to keep the Earth warm. Without the presence of these greenhouse gases, life on Earth could not exist. Figure 1.1 illustrates this natural greenhouse effect.

The Earth is heated by the Sun. It emits heat energy that passes through the atmosphere and heats the Earth. When the Earth heats up, it also emits heat energy. But some of this heat can now be absorbed by the greenhouse gases in the atmosphere, instead of escaping back into space. This keeps the Earth warm and is called the natural greenhouse effect. This is because **these gases act just like the panes of glass in a greenhouse; they allow the Sun's heat in but they trap some of the heat trying to get out.** So the greenhouse becomes hotter than the air outside.

1

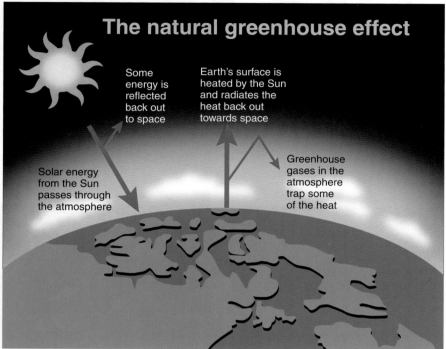

Figure 1.1
The natural greenhouse effect

Climate change

The term **'climate change' means a significant change in global weather patterns over a long period of time**. Figure 1.2 shows how temperatures have changed in the last 100 years. You can see that temperatures have been rising since 1975. Figure 1.3 shows how temperatures have changed in the last 10,000 years. You can see that they have changed even more. At times we have been a little warmer and

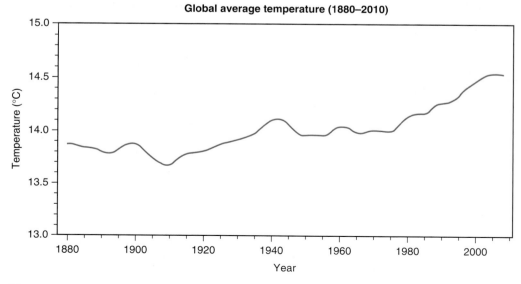

Figure 1.2
Temperature change over the last 100 years

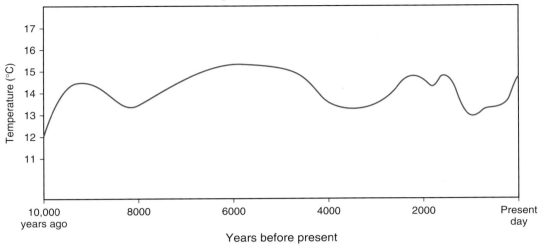

Figure 1.3
Temperature change over the last 10,000 years

10,000 years ago much colder. In fact, climate change has been taking place ever since the Earth was formed and it will surely change just as often in the future.

Because climate change has been going on for so long, we know that human beings cannot have been responsible for all of it. So **climate change must be due to both natural causes and human activities**.

Every country in the world has experienced some form of climate change over the last 10,000 years. We know this because records have been collected from a number of different sources, highlighting both long-term, medium-term and short-term changes. Table 1.1 below describes some of these changes.

Long-term changes *(over many thousands of years)*	Medium-term changes *(over many hundreds of years)*	Short-term changes *(over the last 100–200 years)*
Ice core analysis This is a very effective way of collecting long-term records of temperature changes. In polar regions snow falls every year and never melts. Over time, the layers of snow compact under their own weight to form ice. By drilling through that ice and taking samples of it, records of temperature and atmospheric gases can be built up for periods of hundreds of thousands of years. The samples that are taken contain dust, ash, gas and air bubbles and radioactive substances. These can then be analysed to build a climate record.	**Tree ring analysis** Trees tend to make one growth ring each year, with the newest ring nearest the bark. A year-by-year record or ring pattern is formed that reflects the climate conditions in which the tree grew. For example, a wide ring shows that it was warm and wet because the tree had adequate moisture and a long growing season, while a dry or cold year would result in a very narrow ring.	**Recording changes in biodiversity** Different species of animals and plants are adapted to certain environments. If the conditions change then they have three options: move, adapt or die. By recording changes to the numbers and types of animals and plants in different areas, we can tell how the climate is changing there.

continued

Long-term changes (over many thousands of years)	Medium-term changes (over many hundreds of years)	Short-term changes (over the last 100–200 years)
	Glaciers Glaciers are also a good source of information on climate change. Glaciers are moving rivers of ice and they flow downhill due to gravity. They are very sensitive to temperature fluctuations. If temperatures are consistently low, the glacier will grow and advance; if there is an increase in temperatures the glacier will retreat (melt). It will leave evidence of where it used to be.	**Ice extent** Keeping an annual record of the extent of ice around the world is a good way of showing changes to our climate.
		Measuring air and sea temperatures Making a continuous record of the temperature is a good way of showing whether our climate is changing. By recording temperatures all the time we can compare them with previous years. We have been doing this for over 100 years.

Table 1.1 Examples of measuring long-term, medium-term and short-term climate changes *continued*

National 4

1. Using the word bank below, copy and complete the paragraph, which explains how our atmosphere helps to keep the Earth warm.
   The Sun emits heat energy that passes through the Earth's _____. The Earth heats up and some of its heat energy is radiated back into the atmosphere. The greenhouse gases in the atmosphere _____ this heat energy and store it in the atmosphere rather than letting it _____ into space. This is called the _____ . The main greenhouse gases are _____, _____ and _____.

   carbon dioxide   natural greenhouse effect   methane   absorb
   atmosphere   water vapour   escape

2. Draw a similar diagram to Figure 1.1 in your jotter and label it with the following information:
   - Atmosphere
   - Greenhouse gases
   - Arrow showing the Sun's radiation
   - Arrow showing heat energy from the Earth

3. What does the term 'climate change' mean?

4. Look at Figure 1.2. What was the average temperature in (a) 1880, (b) 2010?

5. Look at Figure 1.3. What was the average temperature (a) 10,000 years ago, (b) 6000 years ago, (c) 2000 years ago?

6. We have sources of evidence that our climate has changed. Explain how we know that there have been changes to our climate:
   (a) in the short term
   (b) in the medium or long term.

National 5

1. Describe, in detail, how the atmosphere helps to keep the Earth warm enough to sustain life.
2. Draw a diagram similar to Figure1.1 and annotate it in your own words.
3. Give a definition of the term 'climate change'.
4. Look at Figure 1.2. Describe the changes in the world's average temperature since 1880.
5. Look at Figure 1.3. When were the coldest and warmest periods in the last 10,000 years and how much colder and warmer were they?
6. Choose one of each of the sources showing how our climate has changed and describe it in detail.

Activities

Activity A

Look at the pictures below and decide which methods of recording climate change are being shown.

A

B

C

Activities continued...

Activity B

Look at the pictures below.

(a) Which picture shows a very cold, dry climate? Why?
(b) Which picture shows a climate which has changed a lot? Why?
(c) Which picture shows a warm, wet climate? Why?

A   B   C

Activity C

Scientists have been keeping a record of average air temperatures and average sea temperatures in country A. The findings for the last ten years are shown below.

Draw a multiple line graph to show the results.

	2003	2004	2005	2006	2007	2008	2009	2010	2011	2012
Air (°C)	10	12	14	11	11	9	12	15	13	11
Sea (°C)	1	1.5	2	1	1	0	1.5	2	1.5	1

 Now complete the 'I can do' boxes for this chapter.

National 5 continued...

- Cold ocean currents replace this warm water by transferring cold water from the poles to the tropics.

3. Describe, in detail, how plate tectonics affects the climate of countries.

Everybody knows volcanoes are hot when they erupt – very, very hot! How can you say that volcanic eruptions make the climate colder?!

4. Explain to this person why he is wrong. Use a diagram to help.
5. The text mentions three things that have either started or stopped an Ice Age. Describe two of these.

Activities

Physical (natural event)		Hotter or colder climate?	Short-term or long-term change?
The Earth's tilt on its axis increases to 25°			
The North Pacific Drift suddenly strengthens			
The Earth's orbit takes it further from the Sun			
All the continents move nearer the equator			
The supervolcano at Yellowstone erupts			
The Brazilian and Mozambique ocean currents stop flowing			
The African crustal plate breaks apart			
Many sunspots form on the surface of the Sun			

Activities continued...

Activity A

Look at the events in the table above. For each one, decide (a) whether it will make the Earth colder or warmer and (b) whether it will be a short-term change (take place over a few years) or a long-term change (take place over thousands of years). Then draw the table and complete it.

Activity B

Choose one of the factors that cause climate change and draw a poster to show exactly how it does so.

Activity C

Using Figure 2.2 and an atlas answer the following questions:

(a) Which ocean current moves warm water from the Gulf of Mexico to Western Europe?
(b) Name one country affected by the Benguela current.
(c) Is the Canaries current warm or cold?
(d) Name two countries in West Africa that are affected by the Canaries current.
(e) Which current passes by Chile?
(f) Which current passes by California?

Now complete the 'I can do' boxes for this chapter.

This chapter looks at the human factors causing climate change.

By the end of this chapter, you should be able to:

✓ describe the difference between the natural and the enhanced greenhouse effect
✓ give examples of some human activities that contribute to climate change
✓ explain how these activities affect the climate.

Climate change – human factors

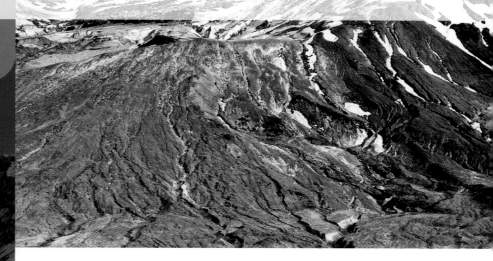

Did you know...?
Carbon dioxide can stay in the atmosphere for 200 years! Nitrous oxide will persist for about 114 years. Methane remains there for about 12 years.

The natural and the enhanced greenhouse effect

In Chapter 1 we looked at the importance of the natural greenhouse effect. This is the effect of having small amounts of greenhouse gases in the atmosphere which allow life on Earth to exist because they trap heat and keep us warm. However, if these gases increase in the atmosphere, they can cause global temperatures to increase.

The enhanced greenhouse effect is caused when human activities release greenhouse gases into the atmosphere (Figure 3.1). When greenhouse gases are in the atmosphere they absorb heat energy, trapping it and causing temperatures to increase. Greenhouse gases can remain in the atmosphere for many years.

Human causes of climate change

Most scientists agree that before the Industrial Revolution people did not affect the climate. Any changes to the climate

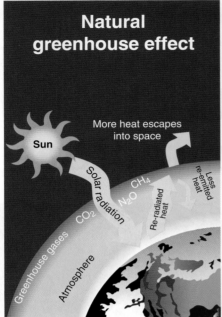

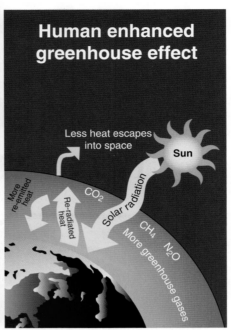

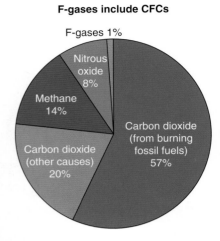

Figure 3.2
Global greenhouse gas emissions

Figure 3.1
The enhanced greenhouse effect

before 200 years ago have to be explained by natural causes. However, since then, we know that **human activities have released a lot of greenhouses gases into the atmosphere, chiefly carbon dioxide, methane, nitrous oxide and chlorofluorocarbons**. And if we have increased the amount of greenhouse gases in the atmosphere, it is highly possible that we have been responsible for changing the climate. Remember we found out in Chapter 1 that, in the last 100 years, global temperatures have increased by nearly 1 °C. Climatologists also predict that by 2100, temperatures could rise by another 1–6 °C if greenhouse gases are not reduced. So what have we been doing to cause the enhanced greenhouse effect?

Increased carbon dioxide (CO_2)

The **burning of fossil fuels such as coal** has resulted in huge amounts of carbon dioxide being released into the atmosphere. Before the invention of electricity coal was very important as it was used to power factories and trains and to warm our homes. Although widespread use of coal in the developed world is less common today, fossil fuels are still burned in power stations to produce electricity.

There are **more vehicles** on the world's roads today than there have ever been. Over one billion cars alone were recorded in 2011. Most vehicles run on petrol, which is a fossil fuel. When it is burned it releases CO_2 into the atmosphere.

Deforestation is an increasing problem across the world, particularly in the rainforests. Trees are known as 'the lungs of our planet' as they take in carbon dioxide and release oxygen. Destroying large areas of forest each year means there are fewer trees to take in carbon dioxide, which means more carbon dioxide remains in the atmosphere.

Figure 3.3
Burning coal increases CO_2 in the atmosphere

Increased methane (CH_4)

Each year 200 million tonnes of methane is released into the atmosphere. Most of the increase in methane comes from **padi fields**, which alone contribute about 10% of all greenhouse gas emissions. Padi fields are flooded fields that are used to grow rice. They are most common in eastern Asia where rice is the main food crop and used to feed millions of people. Rice absorbs carbon from the atmosphere; however, if the plant is not able to use the carbon it is dispersed into the soil where it converts to methane. It is then released into the atmosphere.

Animal dung and belching cows are major contributors to climate change. On average one cow releases about 100 kg of methane each year. Considering there are at least 1.5 billion cows on the planet, that is a lot of methane being released into the atmosphere. Much of the deforestation in the tropical rainforests has taken place in order to increase food production and much of the deforested land is being used for large-scale cattle ranches.

Landfill sites are huge areas of land that have been dug out so that domestic and industrial waste can be buried. As the waste begins to decompose it produces methane, which then goes into our atmosphere.

Increased nitrous oxide (N_2O)

Nitrous oxide is 200–300 times more effective in trapping heat than carbon dioxide and it has one of the longest atmosphere lifetimes of all the greenhouse gasses, lasting for up to 120 years. Since the Industrial Revolution, the level of nitrous oxide in the atmosphere has increased by 16%.

Did you know...?
Each year 6 billion tonnes of carbon dioxide is released into the atmosphere.

Did you know...?
The United Nations believes that 13 million hectares of forest is cut down each year. That is the equivalent to 26 million football pitches!

Nitrous oxide is released when people add nitrogen to the soil by using **fertilisers. Soils release over half of the nitrous oxide in the atmosphere: two to four million tonnes per year.**

Vehicles are also a major contributor to increased levels of nitrous oxide, as it is released when petrol and diesel are burned.

Increased chlorofluorocarbons (CFCs)

Chlorofluorocarbons (CFCs) are chemical compounds that were developed in the 1930s for use in **refrigeration** and **aerosols**. They are also used for **air-conditioning systems** and **polystyrene packaging**. The use of CFCs has been linked to the reduction of the ozone layer but the compounds found in CFCs are also greenhouse gases, which can be more harmful than carbon dioxide.

Did you know...? Deforestation accounts for 20% of all carbon dioxide emissions. This is more than the carbon dioxide emissions from all air travel.

National 4

1. What is the enhanced greenhouse effect?
2. Until 200 years ago, what caused all the changes to the world's climate?
3. Look at Figure 3.2.
   (a) What percentage of all the greenhouse gas emissions is carbon dioxide?
   (b) What are the other main greenhouse gases?
4. What human activities increase carbon dioxide in the atmosphere?
5. Three human activities are increasing methane in the atmosphere. Choose one and describe it.
6. What causes most of the extra nitrous oxides that go into our atmosphere?
7. Why is nitrous oxide a particularly bad greenhouse gas?
8. What are CFCs and where are they found?

National 5

1. What is the difference between the natural greenhouse effect and the enhanced greenhouse effect?
2. Look at Figure 3.2 and describe the relative importance of the greenhouse gases produced by human activity.
3. Describe the human activities which have increased the carbon dioxide in the atmosphere.
4. Which is responsible for more greenhouse gases – crop farming or animal farming? Explain your choice.
5. Describe CFCs and explain where they are found.

Activities

Activity A

Look carefully at the pictures below and for each state:

(a) what the activity is
(b) what gas/gases the activity is releasing into the atmosphere.

A

B

C

D

E

F

Activities continued...

G

H

I

Activity B

Draw a table similar to the one below. List all the main human activities responsible for the enhanced greenhouse effect. Next to each one, write down whether you think the activity is increasing or decreasing, and why. The first activity has been completed for you.

Human activity causing extra greenhouse gases	Increasing or decreasing
Burning fossil fuels	

Now complete the 'I can do' boxes for this chapter.

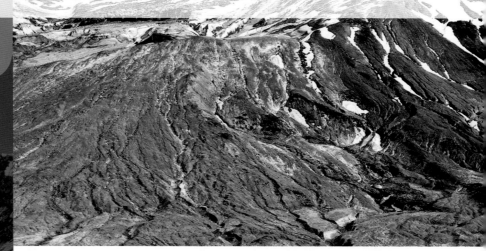

Chapter 4

Climate change – its effects

This chapter looks at the positive and negative effects of climate change.

By the end of this chapter, you should be able to:

✓ give examples of some of the ways that changes in the climate can positively affect countries
✓ give examples of some of the ways that changes in the climate can negatively affect countries.

Climate change and global warming

Global warming refers to the fact that the world has warmed by nearly 1 °C in the last 100 years. Most scientists believe it will rise by at least 2 °C in the next 100 years. Different regions are warming at different rates.

Climate change refers to the fact that it is not just the temperature which is changing; the amount of rainfall is also changing, leading to droughts and floods, and so is wind speed, leading to stronger storms. Different regions of the world are affected in different ways.

Positive effects of climate change

Whenever climate change is mentioned, people think of its negative effects; however, there are some reasons why climate change can be thought of positively. Figure 4.1 shows some of these more positive effects. You can see that northerly latitudes benefit most. Many of these areas that benefit are in developed countries.

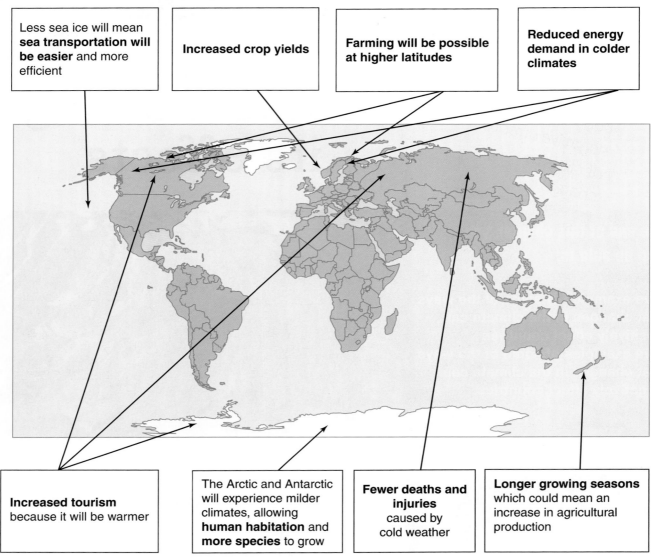

Less sea ice will mean sea transportation will be easier and more efficient

Increased crop yields

Farming will be possible at higher latitudes

Reduced energy demand in colder climates

Increased tourism because it will be warmer

The Arctic and Antarctic will experience milder climates, allowing **human habitation** and **more species** to grow

Fewer deaths and injuries caused by cold weather

Longer growing seasons which could mean an increase in agricultural production

Figure 4.1
Positive effects of climate change

Negative effects of climate change

Some of the negative effects of climate change are already being witnessed across the world, including floods, severe storms, droughts and food shortages. Tropical latitudes are particularly affected and most of these areas are in developing countries. Some of the negative effects of climate change are shown in Figure 4.2.

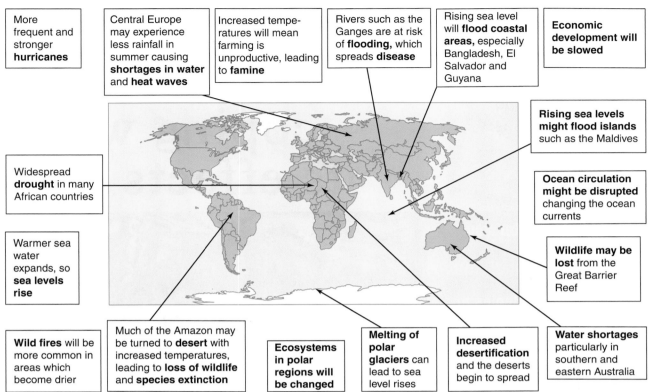

More frequent and stronger **hurricanes**

Central Europe may experience less rainfall in summer causing **shortages in water** and **heat waves**

Increased temperatures will mean farming is unproductive, leading to **famine**

Rivers such as the Ganges are at risk of **flooding,** which spreads **disease**

Rising sea level will **flood coastal areas,** especially Bangladesh, El Salvador and Guyana

Economic development will be slowed

Rising sea levels might flood islands such as the Maldives

Widespread **drought** in many African countries

Ocean circulation might be disrupted changing the ocean currents

Warmer sea water expands, so **sea levels rise**

Wildlife may be lost from the Great Barrier Reef

Wild fires will be more common in areas which become drier

Much of the Amazon may be turned to **desert** with increased temperatures, leading to **loss of wildlife** and **species extinction**

Ecosystems in polar regions will be changed

Melting of polar glaciers can lead to sea level rises

Increased desertification and the deserts begin to spread

Water shortages particularly in southern and eastern Australia

Figure 4.2
Negative effects of climate change

Activities

Write a report on the global effects of climate change.

(a) Begin by briefly describing how the climate is changing and how it will change in the future.

(b) Then, describe and explain what you think are (i) the **three** biggest positive effects, and (ii) the **three** biggest negative effects.

(c) Try and expand on the points made in Figures 4.1 and 4.2. For example, if an area suffers from a severe drought, you can go on to say that crops and animals will die, there will not be enough water, people will have to move away, cities will grow in size, etc.

(d) Make sure you mention specific regions and countries and try and find out more details about them.

(e) Include diagrams and maps in your report.

Now complete the 'I can do' boxes for this chapter.

Chapter 5

This chapter looks at how climate change can be managed.

Climate change – coping with its effects

By the end of this chapter, you should be able to:

✓ give examples of strategies to reduce the effects of climate change at a local level
✓ describe some strategies that can be used on a national level to reduce the effects of climate change
✓ give examples of strategies to reduce the effects of climate change on an international level.

Dealing with climate change

All scientists agree that global temperatures will continue to rise over the next 50 years at least. And this means there will be more floods, droughts, hurricanes, storms, etc. Nearly every country in the world has helped to cause climate change and nearly every country is affected by it. All countries must therefore take responsibility for reducing their harmful effects and every person should play his or her part in helping to reduce climate change.

What can be done at a local level?

You may think that your activities and lifestyle have a tiny effect on climate change in comparison to large-scale activities such as cutting down forests and burning coal in power stations; in fact the choices we make play a major role in climate change. If we are to combat climate change, changes have to be made at all levels, starting with the individual. Here are some of the **things that individuals can do to reduce climate change**.

1. **Reduce, Reuse, Recycle.** Recycling everyday items such as newspapers and milk containers as well as composting unused and wasted food reduces the amount that is sent to landfill sites and therefore reduces greenhouse gas emissions.
2. **Use less hot water.** It takes a lot of energy to heat water; reducing the settings on dishwashers and washing machines will reduce the amount of energy needed.
3. **Insulate your home.** A lot of energy escapes through the roofs of homes that have not been well insulated, so people keep their central heating on for longer or use it more throughout the day. Central heating systems burn fossil fuels; adding more efficient insulation to your home will reduce the need to use central heating and therefore reduce greenhouse gas emissions.
4. **Use energy-efficient light bulbs.** Energy-efficient light bulbs use 75% less energy than regular bulbs.
5. **Turn electrical equipment 'Off'.** It is very common for people to put the TV and other electrical equipment on standby rather than actually turning it off. By using standby the equipment is still using a great deal of energy.
6. **Turn lights off.** If you are not using a room in your home, turn the lights off.
7. **Leave the car at home.** Use public transport where possible rather than taking your car, or walk or cycle short distances.
8. **Cycle to school/work.**

If enough people decide they want to change their habits, companies and businesses will take notice. They will start producing and selling more energy-efficient products; company research departments as well as universities will start inventing new products to meet demand. The government will help to fund this research because it knows it will win votes. Small actions can therefore have a big effect.

What can be done on a national level?

The **UK government** needs everyone to take action themselves, but it also knows it can and must take action as well. As a result, the **government has devised a set of climate change policies**, as shown in Table 5.1.

Did you know....? Transport is the second largest source of carbon dioxide emissions in the UK, accounting for 28% of all emissions.

UK government policy	What does it mean?
The 2050 Challenge	The UK is committed to reducing greenhouse gas emissions by 80% by 2050.
Carbon Budgets	This is a restriction on the total amount of greenhouse gases the UK can emit over a five-year period.
Green Deal	This makes it easier for householders and businesses to pay for some or all of the cost of energy-saving improvements to their properties over time.

continued

CRC Energy Efficiency Scheme	This aims to improve energy efficiency and cut gas emissions in large organisations by requiring them to buy 'allowances' for every tonne of carbon they emit.
Encourage the use of ultra-low emission vehicles	The government provides grants to those who purchase electric, plug-in hybrid and hydrogen-powered cars and vans. It also provides funding to the Plugged-in Places programme.
Energy Bill	This outlines the UK's commitment to increasing the supply of renewable energy and aims to triple the use of renewable electricity by 2015.
Charge on single-use carrier bags	This could lead to a 90% reduction in carrier bags with 12 billion fewer plastic bags in circulation.
Carbon Emissions Reduction Target	Energy companies are obliged to give their customers better deals for being energy efficient.
Tax on 'gas guzzlers'	Cars that emit very low levels of carbon dioxide will pay no road tax; the more carbon dioxide cars emit, the greater the road tax levied on them.
Renewable Transport Fuel Obligation	Encourages the use of sustainable biofuels.

Table 5.1 UK government climate change policies *continued*

What can be done on an international level?

Climate change is global. Countries which cause global warming by burning large amounts of fossil fuels are not the only ones affected by it – the effects are global. International action to tackle climate change is therefore also needed. Two **examples of international strategies** are described below.

United Nations Framework Convention on Climate Change (UNFCCC)

This organisation came up with the Kyoto Protocol in 1997. The treaty sets out binding agreements between some of the most developed nations to reduce greenhouse gas emissions. Each country that signed the protocol agreed their own specific target but, combined, they committed to cut their emissions by 5% between 2008 and 2012. Developing countries do not have binding targets under the Kyoto Protocol but are still expected to reduce their emissions under the treaty. In 2012, the Doha Amendment replaced the Kyoto Protocol.

European Union legislation

The EU has committed to cutting its emissions by 20% by 2020. There are a number of EU initiatives to reduce greenhouse gas emissions, including the European Climate Change Programme (ECCP) and the EU Emissions Trading System. They agreed that by 2020 at least 20% of all the energy used in the EU would be from renewable energy sources. They also made targets to reduce carbon dioxide emissions from new cars and vans, and to support carbon capture and storage technologies.

Did you know....?
The six largest emitters of carbon dioxide (USA, China, Europe, Russia, Japan and India) accounted for 70% of all energy-related carbon dioxide emissions in 2005.

National 4

1. Choose three things that individuals can do to combat climate change and state why they are important.
2. Choose two of the UK's climate change policies that you think will be most successful and describe what they aim to do.
3. Describe the Kyoto Protocol.
4. Why do you think developing countries do not have targets under the Kyoto Protocol?
5. Different countries have different targets to reach for reducing greenhouse gas emissions.
   (a) What is the UK's target for 2050?
   (b) What is the EU's target for 2020?
   (c) What was the UN's target for 2012?
6. Why do individuals have a responsibility to reduce the effects of climate change?
7. Which are likely to have the biggest effect on climate change – local actions, national actions or international actions? Give reasons for your answer.

National 5

1. Choose three things that individuals can do to combat climate change and explain, in detail, why they are important.
2. Choose two of the UK's climate change policies that you think will be most successful and describe, in detail, what they aim to do.
3. Describe the purpose of the Kyoto Protocol and the Doha Amendment.
4. Why do you think developing countries do not have binding targets under the Kyoto Protocol (and the Doha Amendment)?
5. In your own words, describe in detail why it is the responsibility of each individual to reduce the effects of climate change.
6. Different countries have different targets to reach for reducing greenhouse gas emissions. What are the UK's national and international targets?
7. Which are likely to have the biggest effect on climate change – local, national or international actions? Give reasons for your answer.

Activities

Activity A

Design a leaflet to be posted to every household in the UK, showing how greenhouse gas emissions can be reduced. Your leaflet needs to be striking so people read it and don't just throw it away. It also needs to be powerful so people take action as a result.

Activity B

1. Eight local actions are listed in this chapter. Write down which ones will cost you (and your family) money and which ones will save you money.
2. Ten UK climate change policies are listed in Table 5.1. Which policies will cost the government money and which policy will cost the most?

Now complete the 'I can do' boxes for this chapter.

Chapter 6

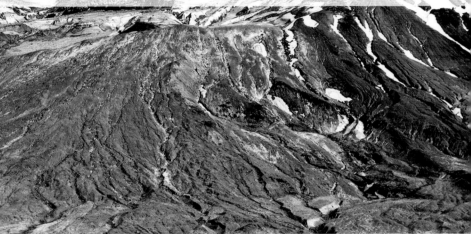

Climate change – case study of Bangladesh

This chapter looks at the effects of climate change on a developing country – Bangladesh.

By the end of this chapter, you should be able to:

✓ describe the human and physical geography of Bangladesh
✓ explain some effects of climate change on Bangladesh
✓ give reasons why Bangladesh finds it difficult to deal with the impact of climate change.

Figure 6.1
The country of Bangladesh

Human geography of Bangladesh

Population	150 million
Population density	1000 people per km²
GNI per capita	$1940
Human Development Index position	140th out of 177 countries
% of people employed in agriculture	66%
Rural/urban population	75% rural/25% urban
% people malnourished	30%
Capital city	Dhaka

Table 6.1
Human geography of Bangladesh

Physical geography of Bangladesh

Bangladesh is a small country in South Asia. It has a land area of 147,000 km², of which 80% is flat floodplain. Three very large rivers flow through Bangladesh: the Ganges, Brahmaputra and the Meghna, all of which join together to form the Ganges Delta before emptying into the Bay of Bengal.

Bangladesh is a very low-lying country with much of it at or near sea level. In fact, 90% of the whole country is less than 10 metres above sea level. Its only hills are in the southeast and northeast.

Bangladesh typically has three seasons: mild winters between October and March; hot, humid summers between March and June; and a warm, rainy monsoon season between June and October. It is one of the rainiest countries in the world and can suffer extreme river and sea floods. Bangladesh also suffers from hurricanes (cyclones) that affect the country between May and November, as well as occasional severe droughts.

Figure 6.2
Typical farming landscape in Bangladesh

Effects of climate change on Bangladesh

Bangladesh is the most vulnerable country to climate change in the world. Over recent years the effects of global climate change have been very obvious here. Some of the effects that the country has witnessed are described below.

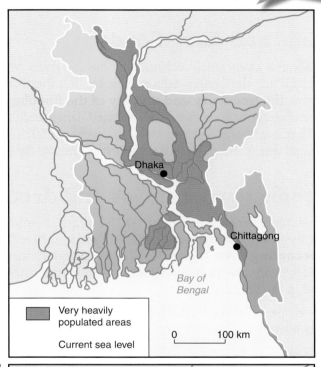

Rises in sea level

Sea levels around the world are rising every year because of polar ice melting (in particular the Greenland ice sheet) and warm water expanding. In 2000, the World Bank estimated that **Bangladesh would see sea level rises of 10 cm by 2020**. By the end of the twenty-first century it is expected that sea level rises will reach at least one metre in Bangladesh. The devastating effects of sea level rises on Bangladesh are shown in Figure 6.3.

Increase in hurricanes

It is thought that hurricane (cyclone) activity has increased globally over the past century. Some scientists say that the number of hurricanes

Did you know....? The sea level along the coast of Bangladesh is rising by about 3 mm each year.

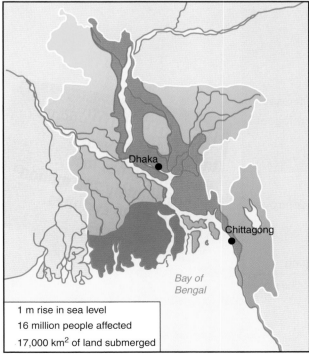

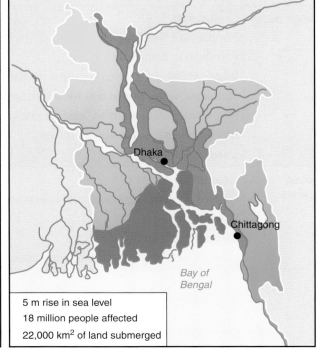

Figure 6.3 Effects of sea level rise on Bangladesh

each year has doubled. This substantial increase in hurricanes is due to rising sea temperatures. Not only have hurricanes become more frequent but they are also increasing in intensity.

Bangladesh now averages sixteen cyclones in every ten years. Five of the ten worst cyclones in history have been in Bangladesh. In 2007, Cyclone Sidr killed 6000 people as it ripped through the coastal areas of Bangladesh. The following year Cyclone Aila killed a further 8000 people. Both Cyclone Sidr and Cyclone Aila are classed as 'super-cyclones' and to have two in two years is very unusual.

Floods and flash floods

Bangladesh is a very small country but it has 250 rivers which flow over almost flat floodplains into the Ganges delta and the sea. It is therefore very vulnerable to floods. However, **the frequency and severity of the flooding is increasing**. In the nineteenth century, six major floods affected Bangladesh; in the twentieth century there were 18 major floods. Severe floods can cover over 60% of the country and remain there for several weeks. Bangladesh is the worst country in the world for flooding.

Extreme temperatures and drought

Temperatures here can reach over 40 °C which, combined with the very high humidity in the rainy season, become almost unbearable. But now **temperatures are becoming even higher**, especially inland in northern and western parts. In addition, every few years Bangladesh suffers from drought which can quickly cause famine in the countryside.

Table 6.2 provides a summary of the problems and effects of climate change on Bangladesh.

Problem	Effects
Sea level rises	Land is lost
	People lose their homes
	Farmland is destroyed
	Water shortages due to saltwater invasion
	Increased water-borne and water-related diseases
	Forced migration leads to 'climate change refugees'
	More people living in shanty towns
	Loss of mangrove forest
Increased hurricanes	Coastal devastation
	Houses, communications and businesses destroyed
	Loss of farmland
	Large-scale loss of life
	Cost of recovery prevents country developing

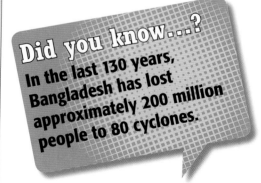
Did you know...?
In the last 130 years, Bangladesh has lost approximately 200 million people to 80 cyclones.

continued

Flooding	People are made homeless
	Loss of life
	Contamination of water leads to disease
	Food supplies are affected as farmland and rice crops are destroyed
	Industries affected
	Communications can be badly damaged
Extreme temperatures and drought	Loss of life due to dehydration
	Famine
	Loss of livestock
	Land is unusable for farming
	Desertification
	Heatwaves cause fatalities

Table 6.2 Impact of climate change on Bangladesh *continued*

Managing climate change in Bangladesh

Although the effects of climate change can be seen the world over, they are most evident in the developing world, in particular Bangladesh. There are two reasons for this: first, the developing world is more dependent on the natural environment, and second, countries in the developing world do not have the means to protect themselves against the growing threat of climate change.

People living in Bangladesh are very heavily dependent on their natural environment through agriculture and fishing. With such a high percentage of people employed in these industries, the effects of climate change impact on the whole country. To try to combat some of the effects of climate change on these industries, Bangladesh has:

- developed and introduced salt-resistant strains of rice
- replaced rice farming with shrimp farming in areas where saltwater has invaded farmland
- given $150 million to 'climate proof' agriculture, for example new sources of freshwater, cyclone shelters and more research into better crop strains.

Bangladesh is one of the poorest countries in the world and it is too poor to introduce sophisticated technologies that countries in the developed world might use. In order to try to fight the effects of climate change Bangladesh desperately needs help from other developed countries, which it believes it is owed as compensation for the global warming caused by these countries.

National 4

1. Look at Table 6.1. How do you know that Bangladesh is a developing country?
2. Give three problems of Bangladesh's physical landscape that makes it vulnerable to the impacts of climate change.
3. Describe the effects of sea level rises on Bangladesh.
4. What is a 'climate change' refugee?
5. Give examples of the effects of cyclones on Bangladesh.
6. Which do you think has the worse effects on the people of Bangladesh – floods or drought? Give reasons for your choice.
7. The sea is gradually taking over farmland in Bangladesh. Describe one way in which the country is coping with this problem.
8. Explain why developing countries such as Bangladesh find it difficult to manage the effects of climate change.
9. Why does Bangladesh believe that other countries should help it manage climate change?

National 5

1. What evidence is there to show that Bangladesh is a developing country?
2. The physical geography of Bangladesh is quite unique. In what ways does the physical landscape make Bangladesh vulnerable to the effects of climate change?
3. Choose the two most serious effects of climate change on Bangladesh and explain how they will affect the country.
4. People in Bangladesh insist that climate change is not in the future; it is happening to them *now*. What is their evidence for this?
5. Explain, in detail, why Bangladesh, like other developing countries, finds it difficult to manage the effects of climate change.
6. Describe the ways in which Bangladesh is coping with its freshwater becoming salty.
7. Do you agree that other countries should help Bangladesh 'as compensation'? Give reasons.

Activities

Activities for this chapter appear at the end of Chapter 7.

Now complete the 'I can do' boxes for this chapter.

Chapter 7

This chapter looks at the effects of climate change on a developed country – Florida, USA.

Climate change – case study of Florida, USA

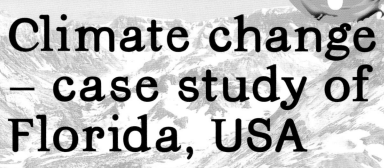
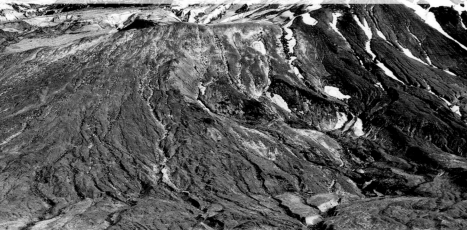

By the end of this chapter, you should be able to:

✓ describe the human and physical geography of Florida
✓ explain some of the effects of climate change on Florida
✓ give reasons why Florida finds it easier to deal with the impacts of climate change.

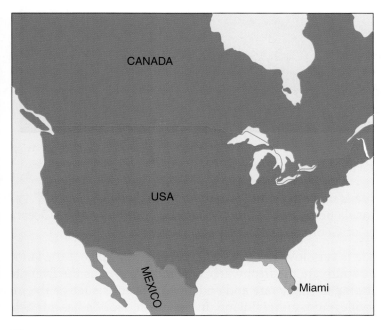

Figure 7.1
The location of Florida

Human geography of Florida

Population	19 million
Population density	353 people per km²
GNI per capita ($)	$48,890 (US average)
Human Development Index position	3rd in the world (USA)
% of people employed in agriculture	3% (US average)
Rural/urban population	21% rural/79% urban (US average)
% people malnourished	N/A
State capital	Tallahassee

Table 7.1
Human geography of Florida

Physical geography of Florida

Figure 7.2
Everglades National Park, Florida

Florida is the most southeasterly state in the USA (Figure 7.1). It is situated on a peninsula between the Gulf of Mexico and the Atlantic Ocean and it borders the states of Georgia and Alabama to the north.

Florida is very low-lying, with approximately 50% of the land below 10 metres. In the south are the freshwater wetlands known as the Everglades (see Figure 7.2) and home to many rare and endangered species such as the manatee, American crocodile and panther. Inland, in the north, Florida has a few hilly areas; however, the highest point is still only 105 metres above sea level.

Florida's climate is divided into two, with a warm temperate climate in the north, which has very warm, long summers and mild winters. The south of the state has a sub-tropical climate with hot and humid summers, a lot of rainfall and mild winters.

The state of Florida suffers natural disasters each year. From June to September Florida is the US state most at risk from hurricanes and it is also one of the most tornado-prone states in the USA.

Effects of climate change on Florida

It is thought that climate change will have the following effects on Florida:

- Temperatures will increase by 2–4 °C over the next 80 years.
- Rainfall will become more intense.
- Droughts will increase, especially in summer.
- Sea levels will rise. **The sea level around Florida's coast is increasing at a rate of approximately 2 cm every ten years.** It is thought that by 2030, the sea level will have increased by approximately 12 cm and by as much as 1 metre by 2100. Eighty per cent of Florida's population lives in the coastal areas where the land is very flat. A 30 cm rise in sea level will move the shoreline inward by more than 300 metres. Figure 7.3 shows estimated sea level rises in Florida.
- **Hurricanes will become stronger.** Florida has seen an increase in intense hurricanes in recent years, such as Hurricane Katrina in 2005 and Superstorm Sandy in 2012. Scientists believe that the number of category 4 and 5 hurricanes will increase by 80% by 2080. They also believe that the frequency of less intense hurricanes (category 1–3) will decrease.

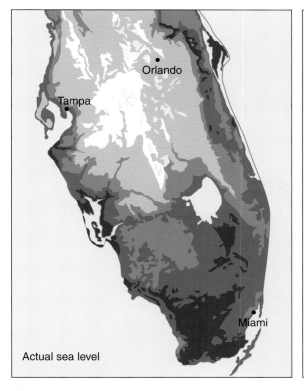

Actual sea level

The result of a five-metre rise in sea level

Figure 7.3
Effects of sea level rises in Florida

Table 7.2 provides a summary of the problems and effects of climate change on Florida.

Problem	Effects
Sea level rise in Florida	Loss of land Loss of mangroves Loss of wildlife and ecosystems Houses will be destroyed Damage to water supply and drainage Saltwater intrusion Increased coastal erosion
Sea level rise in the Everglades	Saltwater intrusion will destroy the Everglades, causing economic losses from tourism of as much as $1 billion Loss of many rare species of plants and animals
Increased hurricanes and storm surges	Coastal devastation Destruction of homes and businesses Loss of life Damage to economy Fewer tourists
More intense rainfall	Widespread flooding Damage to property Damage to businesses/economy Loss of life Loss of crops
Drought in summer	Possible fatalities Loss of crops/farmland Water shortages

Table 7.2
Impact of climate change on Florida

How climate change is managed in Florida

The developed world is in a far better position than developing countries to deal with the effects of climate change. Although developed countries are significantly affected by changes in the world's climate, they have the money to protect themselves, which puts them in a better position. Florida is witnessing the effects of climate change now, and it has put several measures in place in order to combat these effects:

- Raising low-lying roads
- Adding protective beach dunes (see Figure 7.4)
- Obtaining underground water from further inland
- Increasing the use of public transport
- Encouraging the use of solar power and other green (environmentally friendly) energies

- Raising sea walls
- Storing more storm water to supplement drinking water supplies
- Planting more trees in urban areas
- Protecting farmland and open spaces from development

Figure 7.4
Protecting sand dunes in Florida

National 4

1. Look at Table 7.1. How do you know that Florida is a developed country?
2. In what ways does Florida's physical landscape make it vulnerable to the effects of climate change?
3. Describe the problems caused by sea level rises in Florida.
4. Explain how climate change will affect tourism in Florida.
5. Which areas of Florida will be affected by climate change? Give reasons for your answer.
6. Florida is trying to reduce the effects of climate change. Nine measures for doing so are mentioned in the text. Choose two of these measures and explain how each helps.

National 5

1. What evidence is there to show that Florida is part of a developed country?
2. In what ways does the physical geography of Florida make it vulnerable to the effects of climate change?
3. Describe in detail the effects of rising sea levels on (a) the people and (b) the landscape of Florida.
4. Explain how climate change will affect the tourist industry in Florida.
5. Florida is trying to reduce the effects of climate change. Nine measures for doing so are mentioned in the text. Choose three of these measures and explain in detail how each helps.

Activities

Using all the information in Chapters 6 and 7, design a large information poster comparing the effects of climate change on Bangladesh and Florida. Your poster should include information for each area on:

(a) its location
(b) its physical geography
(c) its wealth
(d) the effects of climate change
(e) the ways it is managing the effects of climate change.

Try to design the poster so that it is easy to compare the two areas. Your poster should also include a range of presentation methods, such as maps, graphs, diagrams and pictures.

Now complete the 'I can do' boxes for this chapter.

Chapter  8

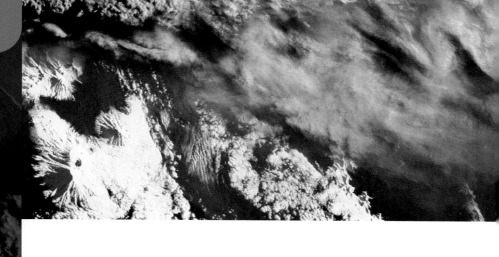

This chapter looks at the structure of the Earth.

Structure of the Earth

By the end of this chapter, you should be able to:

✓ explain the meaning of the term *natural (environmental) hazard*
✓ describe the three main layers of the Earth
✓ describe the two types of the Earth's crust.

Natural or environmental hazards

Natural hazards, also known as environmental hazards, are sudden events in nature that cause people problems. The problems may be slight (e.g. snow blocking roads) or severe (e.g. forest fires destroying property) or catastrophic (e.g. volcanic eruptions, earthquakes, drought, floods and tropical storms, which may kill hundreds of people).

The worst natural hazards are called natural disasters and are thought to kill on average 130,000 people every year, 97% of whom live in developing countries. It is estimated that they cause damage totalling £60 billion a year. This topic looks at the most serious environmental hazards.

Did you know...?
Almost half the people in the world have lived through a natural disaster in the last ten years.

The Earth's structure

The Earth is made up of three layers: the core, the mantle and the crust.

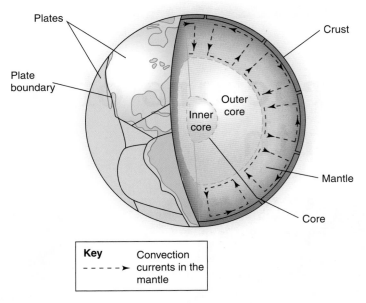

Figure 8.1
Structure of the Earth

At the centre of the Earth is the core, which is made up mostly of nickel and iron, where the temperature is thought to be between 4000 and 6000 °C. Very little is known about the core of the Earth because it is simply too hot.

The layer that surrounds the core is called the mantle. This is the thickest layer. The mantle is made up of molten rock (magma). Temperatures in the mantle can reach 3000 °C. Because the rocks in the mantle are molten or semi-molten they can move. They move because they are so hot. As the rocks heat up, they rise from the core to the crust, then cool down, spread out and sink back to the core (see Figure 8.1).

The crust is the hard outside layer of the Earth. It is the thinnest part of the Earth. If the Earth is compared to an apple, the crust would be as thin as the skin of the apple. The crust 'floats' on top of the mantle.

There are two types of crust: continental crust and oceanic crust. Continental crust is the oldest type of crust and is about 4 billion years old. The continental crust is the layer of rock that forms the continents. It is much thicker, lighter and older than that found under the oceans.

The oceanic crust is thinner, heavier and much younger; its oldest rocks are about 230 million years old, many are only a few million years old and some are only a few years old.

The crust is not one single piece but is broken into many different pieces called tectonic plates. These plates are constantly moving because they float on the molten rock in the mantle, and the molten rock is moving.

Activities

Activity A

Draw the structure of the Earth in your notebook. Colour the layers as follows:

- Core = Red
- Mantle = Orange
- Crust = Green

Activity B

Read through the statements below about the structure of the Earth. Some of these statements are true and some are false. Identify what is wrong with each false statement and re-write it correctly.

1. The Earth is made up of three layers.
2. At the centre of the Earth is the crust. This can be split into two parts, the inner crust and the outer crust.
3. Temperatures at the centre of the Earth can be as high as 10,000°C.
4. The thickest layer is called the mantle.
5. The rocks in the mantle are constantly moving and are called convection currents.
6. If you compare the structure of the Earth to an apple, the crust would be the juicy middle bit.
7. There are two types of crust: continental crust and oceanic crust.
8. Continental crust is much thinner and younger than oceanic crust and is found under the oceans.
9. Oceanic crust is found under the oceans and all of it is about 230 million years old.
10. Continental crust is thicker than the oceanic crust, so it is heavier.

Now complete the 'I can do' boxes for this chapter.

Chapter 9

Crustal plates and plate boundaries

This chapter looks at crustal (tectonic) plates.

By the end of this chapter, you should be able to:

✓ give a definition of a crustal plate
✓ describe the four types of plate boundary
✓ describe the activities which take place at each plate boundary.

Plate boundaries

Figure 9.1 shows that **the crust of the Earth is split into separate blocks called crustal plates**. The area where two plates meet is called a plate boundary. There are four types of plate boundary and different activities take place at each one.

Destructive plate boundaries

Destructive plate boundaries (Figure 9.2) are found where an oceanic plate and a continental plate are moving together. One plate is made of oceanic crust and the other is made of continental crust. Because oceanic crust is heavier it is pushed downwards into the mantle where it melts, immediately forcing magma through the cracks to the Earth's surface and causing a volcanic eruption. Meanwhile, the surface rocks crumple together to form fold mountains. The rocks also crack as they are squeezed up, which triggers earthquakes.

This is taking place in Japan, in Indonesia and along the west coast of South America.

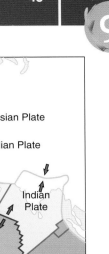

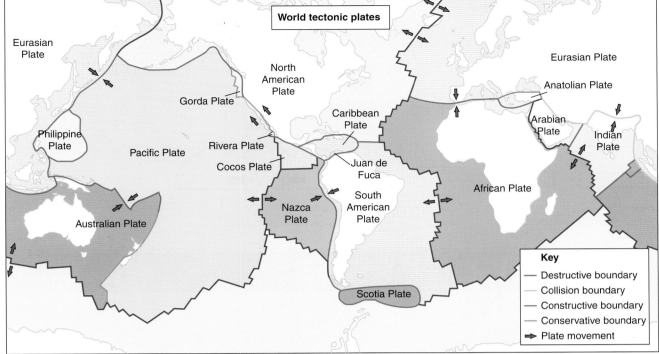

Figure 9.1
Crustal plates and plate boundaries

Constructive plate boundaries

When two plates move apart, they form a constructive plate boundary (Figure 9.3). The crust cracks and splits, which allows magma from the mantle to reach the surface and cause a volcanic eruption. When the magma cools, it solidifies and forms new crust. Earthquakes also occur as the rocks split and move.

This is taking place in the middle of the Atlantic Ocean and the Southern Ocean near Antarctica.

Conservative plate boundaries

When two plates slide past each other they form a conservative plate boundary (Figure 9.4). The sliding movement is not a smooth, continuous process; for most of the time the two plates are locked together. But, as pressure builds up, the plates suddenly jerk past each other causing an earthquake.

This is taking place in California, Turkey and New Zealand.

Collision plate boundaries

Finally, when two continental plates move together, they form a collision plate boundary (Figure 9.5). The crust is too light to sink; instead it is forced upwards as the two plates come together. Over time this forms mountain ranges such as the Himalayas and the Alps. Earthquakes also occur as the rocks stretch and crack.

This is taking place in the Himalayas in Asia and the Alps and Atlas Mountains.

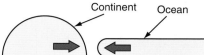

Figure 9.2
Destructive plate boundary

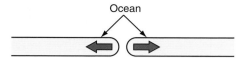

Figure 9.3
Constructive plate boundary

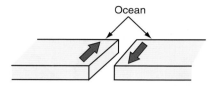

Figure 9.4
Conservative plate boundary

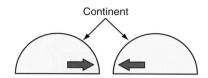

Figure 9.5
Collision plate boundary

National 4

1. What is:
   (a) a crustal plate
   (b) a plate boundary?
2. What is the difference between a destructive plate boundary and a collision plate boundary?
3. Explain how the molten rock is able to reach the surface at a constructive plate boundary.
4. Draw a diagram to show a destructive plate boundary and label it to explain why earthquakes and volcanoes occur there.
5. Name the different plate boundaries where:
   (a) earthquakes occur
   (b) volcanoes occur.
6. Copy and complete Table 9.1 below. You may need to use an atlas to help you.

Type of plate boundary	Description	Activity	Locations	Example
Collision				Himalayas
		Earthquakes Volcanoes Fold mountains	Nazca and South America	Andes
	Two plates move away from each other	Volcanoes Earthquakes		
	Two plates slide past each other			San Andreas Fault, USA

Table 9.1

National 5

1. Describe what is meant by a crustal plate and a plate boundary.
2. In your own words, summarise the movement of plates at each of the different plate boundaries.
3. Draw a diagram to show a destructive plate boundary and label it to explain why earthquakes and volcanoes occur there.
4. Copy and complete Table 9.1. You might need to use an atlas to help you.
5. Explain why fold mountains form at two types of plate boundary.
6. Explain why volcanoes are found at two types of plate boundary.

Activities

Activity A

Using an atlas, Figure 9.1 and the information in this chapter, find out near which type of plate boundary the cities/islands listed below are found. Choose from: destructive, constructive, conservative or collision.

Reykjavik	Ankara	Jakarta	Tokyo
Lhasa	Lima	San Francisco	Christchurch
Algiers	St. Helena (island)	Kerguelen (island)	

Activity B

You will need a blank map of the world to complete this activity.

Scientists believe that all the continents were once joined together in a huge super-continent which they have called Pangaea. There is plenty of evidence for this.

(a) Complete the map using the information shown in the boxes below. You will need to draw a key underneath your map to show the information clearly.
(b) Imagine you are at a meeting where you must convince the audience that the continents of the world were once all joined together. Write your speech describing the evidence that our continents used to be joined together. Make the evidence as convincing as possible. At the end of the speech explain how it is possible for enormous continents to move across the world.

Matching rocks

There are igneous rocks in eastern India which are exactly the same as the igneous rocks in Western Australia. There are rocks containing iron in Western Australia exactly the same as the iron in rocks in South Africa.

Colour the rocks *red* along the:

- east cost of India
- west coast of Australia
- east coast of South Africa.

Activities continued...

Lystrosaurus

Lystrosaurus was a dinosaur that lived about 200 million years ago. It died out long before humans appeared on Earth. It could not swim or fly, however fossils of *Lystrosaurus* have been found in:

- South Africa
- Antarctica
- India.

Colour each of these areas in *green*.

The continents' shapes

The east coast of South America looks as if it fits into the west coast of Africa.

Colour each of these coastlines in *blue*.

Matching mountains

The Appalachian Mountains run along the east coast of North America. The Scottish Highlands are exactly the same age and rock type as the Appalachians. So are the Norwegian Mountains. So are the mountains in Greenland.

Colour each of these mountain ranges in *brown*.

Now complete the 'I can do' boxes for this chapter.

Chapter 10

This chapter looks at volcanoes.

Volcanoes

By the end of this chapter, you should be able to:

✓ describe the location of volcanoes around the world
✓ explain the formation of volcanoes at plate boundaries
✓ describe the features of a volcano.

Volcanoes as natural hazards

Volcanoes find many ways of causing people problems. Volcanic ash can cover houses and streets, lava can pour out over farmland and people may be forced to leave their homes when a volcano erupts. At their worst, volcanoes are killers.

The box below describes some of the worst volcanic eruptions in human history.

Mt. Pelee (1902): 28,000 people killed by a ball of lava that hurtled down the side of the volcano at 300 km/h.

Vesuvius (AD79): 2000 people suffocated by a massive downfall of hot volcanic ash that buried the town of Pompeii to a depth of 3 m in a very short time.

Krakatoa (1883): 36,000 people killed by tsunami up to 35 m high.

Nevada del Ruiz (1985): 20,000 people buried by a 40 m high mudflow (ash mixed with snow melt) sweeping down the volcano at 50 km/h, which then turned solid and trapped them.

Location of volcanoes

Figure 10.1 shows the location and distribution of active volcanoes in the world. **Active volcanoes are those that are likely to erupt**, for example, Mt. Etna. Extinct volcanoes are those that will never erupt again, for example, Edinburgh's volcano. There are also dormant volcanoes that have not erupted for at least 100 years, but may erupt again.

Active volcanoes are concentrated in just a few areas of the world. **Most are found near crustal plate boundaries.** In particular, they are located around the edge of the Pacific Ocean (e.g. Mt. St. Helens, Fujiyama), in the middle of the Atlantic Ocean (e.g. Surtsey) and through the Mediterranean Sea (e.g. Vesuvius, Etna).

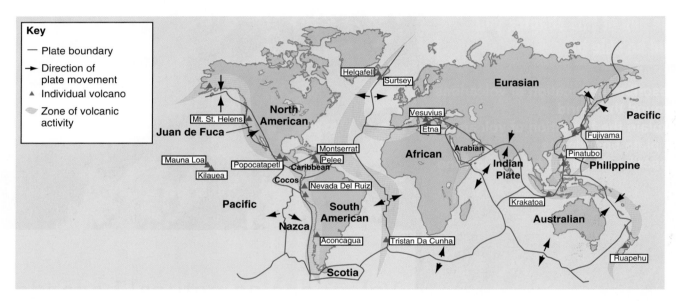

Figure 10.1
Distribution of plate boundaries and volcanoes

Volcanoes at constructive plate boundaries

Constructive plate boundaries are found where two plates are moving apart (see Figure 10.2). As two plates move away from each other, liquid rock from the mantle rises through the cracks. When the magma reaches the surface an eruption will occur. On the surface, the lava begins to cool and becomes solid rock. This fills the crack where the two plates moved apart. As the plates continue to move apart, more and more cracks appear and the process repeats itself. Every time an eruption occurs, a layer of lava and ash is laid down over the land.

Volcanoes at destructive plate boundaries

Destructive plate boundaries are found where two plates move together (Figure 10.3). As two plates move towards each other, the heavier oceanic plate is pushed downwards into the mantle, where it melts and the liquid rock (magma) makes its way through the cracks in the surface and erupts over the land. These eruptions can be very explosive.

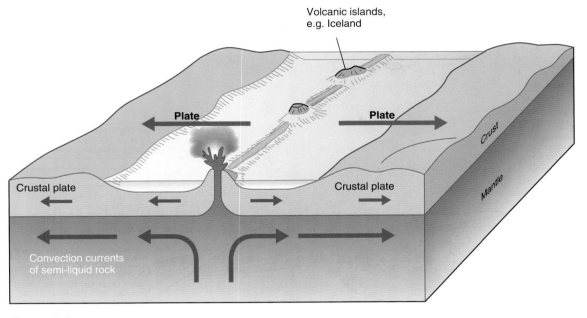

Figure 10.2
Volcanoes at constructive plate boundaries

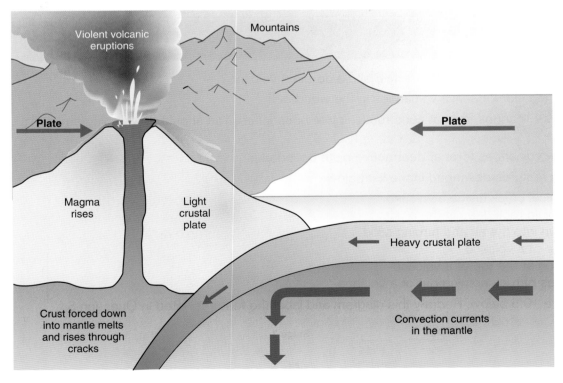

Figure 10.3
Volcanoes at destructive plate boundaries

Features of a volcano

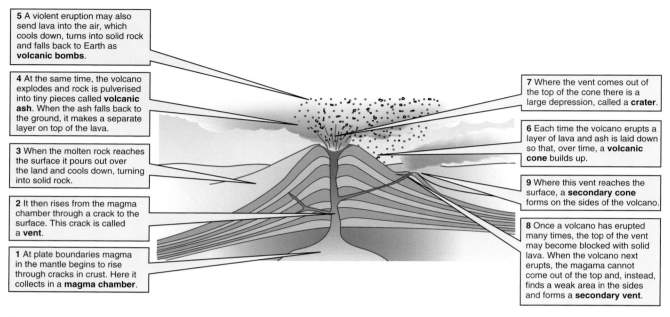

5 A violent eruption may also send lava into the air, which cools down, turns into solid rock and falls back to Earth as **volcanic bombs**.

4 At the same time, the volcano explodes and rock is pulverised into tiny pieces called **volcanic ash**. When the ash falls back to the ground, it makes a separate layer on top of the lava.

3 When the molten rock reaches the surface it pours out over the land and cools down, turning into solid rock.

2 It then rises from the magma chamber through a crack to the surface. This crack is called a **vent**.

1 At plate boundaries magma in the mantle begins to rise through cracks in crust. Here it collects in a **magma chamber**.

7 Where the vent comes out of the top of the cone there is a large depression, called a **crater**.

6 Each time the volcano erupts a layer of lava and ash is laid down so that, over time, a **volcanic cone** builds up.

9 Where this vent reaches the surface, a **secondary cone** forms on the sides of the volcano.

8 Once a volcano has erupted many times, the top of the vent may become blocked with solid lava. When the volcano next erupts, the magama cannot come out of the top and, instead, finds a weak area in the sides and forms a **secondary vent**.

Figure 10.4
Features of a volcano

National 4

1. What is the difference between an active and a dormant volcano?
2. Describe the location of active volcanoes in the world.
3. Describe how volcanoes form at constructive plate boundaries. You should draw a diagram in your answer.
4. Describe how volcanoes form at destructive plate boundaries.
5. Name the six features described in the list below:
   (a) wide vertical crack inside a volcano
   (b) pool of liquid rock deep in the crust
   (c) lava blown into the air and turning solid
   (d) the shape of a volcano
   (e) depression at the top of a volcano
   (f) tiny pieces of rock coming out of a volcano
6. Look at Figure 10.5 below. Redraw this diagram and label the features listed in Question 5.

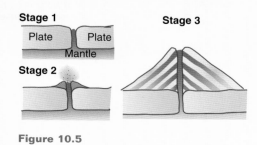

Figure 10.5

National 5

1. Volcanoes are classed using names that describe how recently they erupted. What are these three names and what do they mean?

2. Using the information shown in Figure 10.1, describe, in detail, the location of active volcanoes in the world.

3. Using a diagram, explain why volcanoes are found at destructive plate boundaries.

4. Describe, in detail, how volcanoes form at constructive plate boundaries.

5. Name the six features described in the list below:

   (a) wide vertical crack inside a volcano
   (b) pool of liquid rock deep in the crust
   (c) lava blown into the air and turning solid
   (d) formed when the volcano's main vent is blocked
   (e) depression at the top of a volcano
   (f) made of layers of lava and ash

6. Look at Figure 10.5. Redraw this diagram, label the features shown and add a magma chamber and secondary vent.

Activities

Activity A

Using the information in the box describing notorious volcanic eruptions (on page 47), draw a bar graph showing the number of people that were killed during each volcano's eruption.

Activity B

The following sentences describe how a volcano forms at constructive plate boundaries. However, the information is shown in the wrong order. Rearrange the sentences into the correct order.

- On the surface, the lava begins to cool and becomes solid rock.
- As the two plates move away from each other, liquid rock from the mantle rises through the cracks.
- As the plates continue to move apart, more and more cracks appear and the process repeats itself.
- Every time an eruption occurs, a layer of lava and ash is laid down.
- When the magma reaches the surface an eruption will occur.
- This fills the crack where the two plates moved apart.

Activity C

The following sentences describe how volcanoes form at destructive plate boundaries. Some of the information in *each* sentence is incorrect and the sentences are not in the correct order. Correct the information in each sentence and then rearrange the sentences into the correct order.

Activities continued...

- As it is pushed downwards, it solidifies.
- The solid rock (magma) then makes its way through the cracks in the surface and explodes.
- One of the continental plates is pushed downwards into the core.
- Two continental plates come together.

Activity D

Make your own volcano!

You will need:

- Clay
- Paints
- Plastic bottle
- Warm water
- Red food colouring
- Washing-up liquid
- Baking soda
- Vinegar

Instructions

1. Place the plastic bottle in the middle and mould your clay around it, to form a volcano shape.
2. Colour your clay with paint.
3. Pour the warm water into the plastic bottle and add a couple of drops of red food colouring.
4. Add six drops of washing up liquid to the bottle.
5. Add two tablespoons of baking soda.
6. Slowly pour the vinegar into the bottle and watch your volcano erupt!

Now complete the 'I can do' boxes for this chapter.

Chapter

11

The eruption of Mt. St. Helens, 1980

This chapter looks at the eruption of Mt. St. Helens.

By the end of this chapter, you should be able to:

✓ explain the reason for the eruption
✓ give examples of the effects of the eruption on the landscape
✓ describe the impact of the eruption on the people.

Cause of the eruption

Mt. St. Helens is in the Rocky Mountains near the west coast of the USA, in the state of Washington. **It lies near to a destructive plate boundary** (see Figure 11.1), where the small Juan de Fuca Plate is moving southeast and the North American Plate is moving northwest.

The small plate is being forced under the larger plate and into the mantle. Here it melts, partly because of the heat and partly because of the immense friction as two plates grind together. **As it melts, molten rock rises into the crust.** Here it builds up in magma chambers until it is able to force its way through cracks in the crust to the surface (Figure 11.2). This has happened many times, for example at Mt. Lassen (also known as Lassen Peak) in 1914, Mt. Rainier in 1834 and, catastrophically, at Mt. St. Helens in 1980.

The eruption

On 18 May 1980 Mt. St. Helens erupted for the first time in 123 years. It erupted with a power 500 times greater than any atomic bomb exploded during World War Two and was the

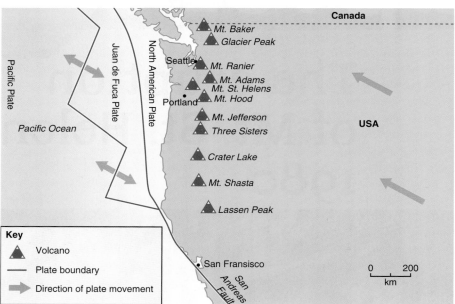

Figure 11.1
The location of Mt. St. Helens

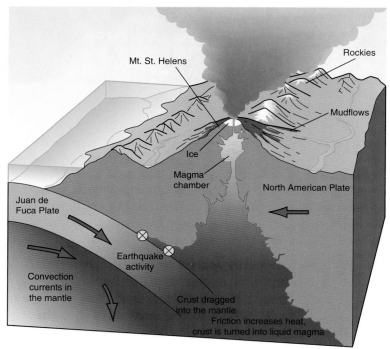

Figure 11.2
Cause of the eruption of Mt. St. Helens

most powerful eruption on Earth in the last 60 years. No lava poured out, but the eruption still had four devastating effects:

1. The eruption was triggered by the **biggest landslide in recorded history**, which sped down the north side of the mountain at 250 km/h.
2. There was a **tremendous blast** from the eruption, which could be heard 300 km away. The blast travelled at 500 km/h but this increased at times to twice that speed, overtaking the landslide. The blast contained rock,

ash and gases at temperatures of over 300 °C, known as a **pyroclastic flow**.

3. A **mudflow of rock, melted ice and ash** hurtled down the mountain side at 250 km/h. The heat from the eruption had melted ice and snow on the mountain, releasing 200,000 million litres of water.

4. Around 400 million tonnes of **ash rose 20 km into the air**. Some rose so high, it never came down.

Did you know....?
Enough trees were flattened by the blast to build 300,000 family homes.

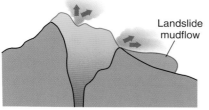
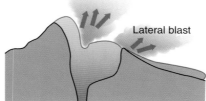

Landslide mudflow

Lateral blast

Ash cloud

Figure 11.3
The eruption of Mt. St. Helens

Impact on the landscape

- The landslide buried the North Fork Toutle River (see Figure 11.5) to a depth of 200 metres.
- The blast and pyroclastic flow killed every form of plant and animal life for a distance of 25 km north of the volcano. Even fully-grown fir trees were flattened, up to 30 km away. About 7000 animals died, including elk and bears.
- The mudflow choked rivers with sediment, killing all fish and water life and completely filling in Spirit Lake. About 12 million salmon died. The mud emptied itself into the sea at Portland, clogging up the harbour.
- The eruption of ash blew away the top of the mountain. In seconds it changed from a mountain 2950 metres high to one that was only 2560 metres high. At the top a crater 500 metres deep formed.

Figure 11.4
Trees flattened by the lava flow

Impact on the people

- The eruption on 18 May 1980 occurred on a Sunday, so no one was working in the forests that cover the slopes of Mt. St. Helens. Local people had been evacuated from their homes and tourists were prevented from getting close. In spite of all this, the eruption still killed 57 people and 198 had to be rescued. Damage ran into billions of dollars.
- **Mt. St. Helens had given clear warnings that it might erupt explosively.** From March onwards there had been minor earthquakes and small eruptions of ash and steam. These gradually became more severe.
- As a result, **the authorities were able to evacuate residents, tourists and forestry workers** from the surrounding area. They researched the area

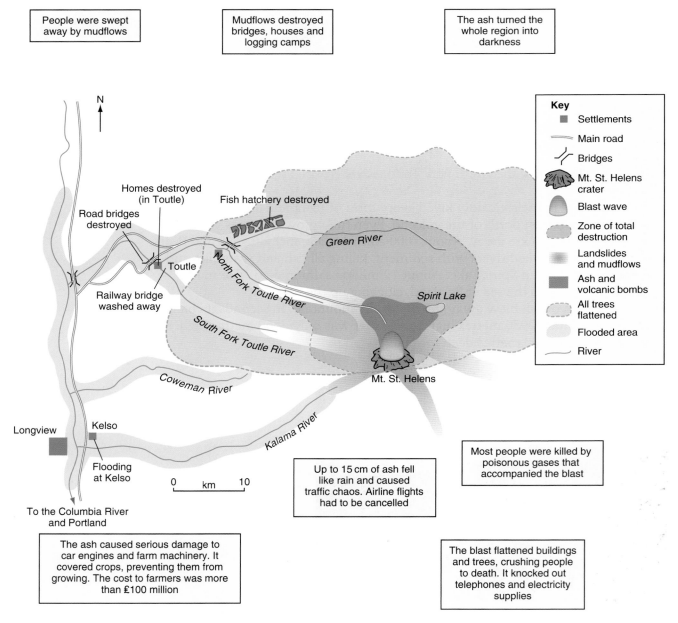

Figure 11.5
Effects of the eruption of Mt. St. Helens

affected by the previous eruption and made this an exclusion zone around Mt. St. Helens. **Emergency services were on hand**, including helicopters and aeroplanes.

■ But the scientists could not give a precise date for the eruption. They tried measuring (a) the frequency of earthquakes on the mountain – the greater the frequency, the nearer the eruption, and (b) the size of the volcanic cone – the volcano bulged as magma built up in the vent. Even the day before the eruption, scientists were stating that the eruption might still be a few weeks away. **Nor did the experts predict that the blast from the eruption would be from the north side.** As a result, 90% of the people killed were outside the exclusion zone.

Figure 11.6
Vehicles trapped by the eruption

National 4

1. Where is Mt. St. Helens?
2. When did the eruption of Mt. St. Helens take place?
3. Explain why the eruption took place.
4. Give three pieces of evidence to show that the eruption was violent.
5. Describe the effects on the landscape of the blast and the mudflow.
6. Which was the worst effect on the local people – the ash eruption, the mudflow or the blast? Give reasons for your answer.
7. Were the authorities able to predict that Mt. St. Helens was going to erupt? Give reasons for your answer.
8. How well did the authorities plan for the eruption?

National 5

1. Describe the location of Mt. St. Helens and the date of its eruption.
2. Explain, in detail, why Mt. St. Helens erupted.
3. Copy the table below and fill in the details by writing the different effects of the eruption in the appropriate box.

	Effects of Mt. St. Helens eruption 1980	
	On the landscape	On the people
Ash eruption		
Blast		
Mudflow		

4. Describe the different methods of predicting the eruption of Mt. St. Helens.
5. Explain how effective these methods were.
6. Do you think the authorities were well prepared for the eruption? Give reasons for your answer.

Activities

Activity A

Imagine the date is 17 May 1980 and you and your friends are arguing about whether to camp on the lower slopes of Mt. St. Helens. What would be the main arguments for and against camping?

Activity B

After the argument you decide to camp on Mt. St. Helens and the next day the volcano erupts. Describe what you see, feel and hear.

Now complete the 'I can do' boxes for this chapter.

Chapter  12

This chapter looks at how the people coped with the eruption of Mt. St. Helens.

Managing the eruption of Mt. St. Helens, 1980

By the end of this chapter, you should be able to:

✓ give examples of help that was given before the eruption
✓ describe some of the short-term aid needed following the eruption
✓ describe what long-term aid is and why it was needed.

After any natural disaster, there is both official and voluntary aid to help the area recover. **Voluntary aid is provided by charities** and individuals; **official aid comes from governments**. Most aid usually comes from the home country but other countries may also send help.

After any natural disaster, there is a need for short-term aid and long-term aid. **Short-term aid is needed to rescue people** and then to provide emergency help for the survivors – food, clean water, medicine, tents and blankets. It is also needed to clear up the area and make it possible for people to survive.

Long-term aid is needed to allow the area to return to normal.

Aid can be in the form of **money**, but it can be **goods** (machinery, medicine, food) and it can also be **skilled people**.

The US federal government gave the most aid altogether, totalling $950 million.

Figure 12.1
Vehicle buried by ash

Figure 12.2
Road block on Mt. St. Helens

- Part of the mountain was closed off before the eruption.
- About 2000 people evacuated before and during the eruption.
- This saved many lives.
- But people were still allowed to live and work near the mountain; and some refused to leave.
- Rescue centres were set up.
- 198 people were rescued.
- 57 people died.

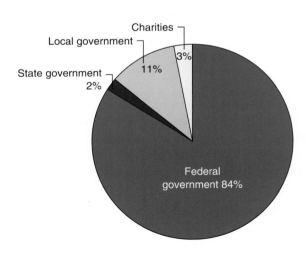

Figure 12.3
Aid given for the clean-up operation after the Mt. St. Helens eruption

Figure 12.4
Removing ash

- 1 million tonnes of ash were removed from roads and airports, at a cost of $1 million.
- 200,000 people were temporarily employed in the clean-up operation.
- But local people were unprepared for the ash fall or its effects, in particular on transportation and water treatment systems.
- Millions of trees were replanted.
- It will be 2050 before all the plants, trees and wildlife return to normal.

Figure 12.5
Replanted trees on Mt. St. Helens

Money was given by the US government to rebuild houses, repair roads and construct a new highway, costing $145 million.

The private National Science Foundation funded 74 research projects on the eruption over ten years, costing $5 million.

Federal and state money was given for new salmon hatcheries and farmers were also compensated because their crops had been ruined by the ash fall.

The Mt. St. Helens Visitor Center was built to attract tourists back to the area, together with more trails and information centres. This cost $50 million of federal money.

Local rivers were dredged to remove logs and levées were built up beside the Columbia River to reduce flooding.

Many charities and local people helped in the relief effort after the eruption.

Activities

Write a report on the different types of aid given for the eruption of Mt. St. Helens in 1980, mentioning how effective each was.

Now complete the 'I can do' boxes for this chapter.

Chapter  13

Earthquakes

This chapter looks at the location and features of earthquakes.

By the end of this chapter, you should be able to:

✓ describe where earthquakes occur
✓ explain how an earthquake happens
✓ describe how earthquakes are measured.

Earthquakes as natural hazards

Each year 20,000 people are killed by earthquakes, which makes them bigger killers than volcanoes. This is partly because earthquakes give no warning. It is also because many areas that suffer earthquakes are popular areas in which to live, such as California, and, in some cases, people do not know they are living in such a dangerous place. Earthquakes are also very common.

An earthquake happens somewhere in the world every two minutes. But most are very slight and they mainly occur under the sea. No one hears of them. Sometimes, however, there are severe earthquakes and, just occasionally, they take place under a large town. This is when earthquakes make headline news.

---

31 May 1970
Earthquake in Peru Causes Landslide
50,000 feared buried alive

18 April 1906
Fires Destroy San Francisco
caused by severe earthquake

11 March 2010
Tsunami Alert
as quake hits Taiwan

12 January 2010
Chaos in Haiti
200,000 feared dead
in massive earthquake

Location of earthquakes

The location of earthquakes (Figure 13.1) is very similar to that of volcanoes. They are concentrated in just a few parts of the world. **Nearly all take place near crustal plate boundaries.**

Earthquakes are particularly common around the edge of the Pacific Ocean (e.g. Japan, California) and through the Mediterranean Sea (e.g. Turkey, Italy). Most earthquakes occur under the sea (e.g. mid-Atlantic) because most plate boundaries are found there.

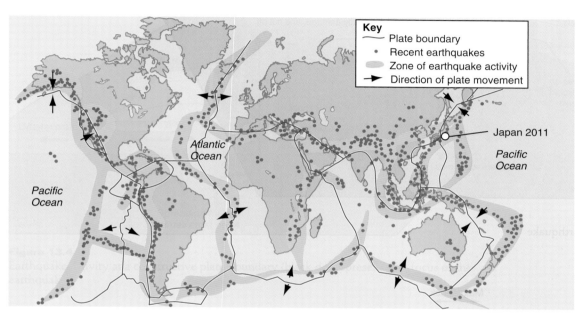

Figure 13.1
Distribution of earthquakes

Features of an earthquake

An earthquake occurs when rocks inside the crust move suddenly. Where this happens is called the *focus* of the earthquake. **This sudden movement causes shock waves** to travel out in all directions. The place on the surface directly above the focus receives the worst shock waves. This is called the *epicentre* (Figure 13.2). **There are three types of shock waves:**

1. **P waves** (push or primary waves) make the rocks move up and down – they travel the fastest.
2. **S waves** (shake or secondary waves) make the rocks move from side to side – they travel at two-thirds the speed of P waves.
3. **L waves** (long waves) spread out in waves along the surface – they are the slowest but the most destructive.

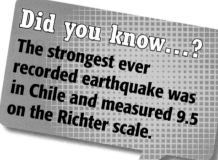

Did you know....?
The strongest ever recorded earthquake was in Chile and measured 9.5 on the Richter scale.

The **shock waves are detected on seismographs or seismometers** (Figure 13.3). The magnitude of the earthquake is **measured on the Richter scale**. This is a logarithmic scale from 1–12. Earthquakes of scale 3 or under are minor and are not usually strong enough to be felt. Scale 6 or more are severe. No earthquake has yet registered a scale 10.

Conservative (sliding) plate boundaries

In some areas of the world **where crustal plates meet, they just slide past one another**. No crust is destroyed or constructed and no volcanic activity takes place. But **the sliding movement is not smooth**. Because of the immense friction between the two slabs of crust, **the plates are locked together most of the time**. When the pressure has built up over a long period of time, it is great enough to overcome the friction and this is when **one plate suddenly jerks past the other**. This causes shock waves and an earthquake on the surface. An example of where this occurs is California.

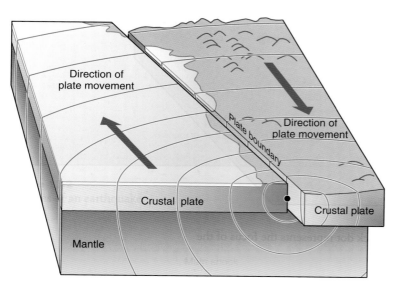

Figure 13.6
Earthquake activity at a conservative (sliding) plate boundary (black dot represents the focus of the earthquake)

National 4

1. Why do earthquakes kill more people than do volcanoes?
2. Why do you think we only hear of the earthquakes that happen under large towns?
3. Describe the location of earthquakes.
4. The words in the list below and their definitions have been jumbled up. Match each word to its correct definition.

Focus	A logarithmic scale showing the magnitude of the earthquake
Epicentre	The instrument used to measure the strength of the earthquake
Seismograph	The point where the rocks move suddenly inside the Earth's crust
Richter scale	The point on the Earth's surface where the earthquake occurred

National 4 continued...

5. Name the three types of shock waves.

6. An earthquake is caused by shock waves at the Earth's surface. What causes the shock waves?

7. Copy and complete the paragraphs below describing how earthquakes occur at the three different plate boundaries. Use the word banks below to help you.

    At constructive plate boundaries, the plates are moving _____ from each other. As this happens, some of the rocks crack and move _____. This causes _____, which travel through the crust to the surface where they cause the ground to _____.

    shake   sharply   away   shock waves

    At destructive plate boundaries, the plates are moving _____ each other. One plate is forced down into the _____ by the other. As the plate moves down, pressure builds up and the plate _____ downwards. This sudden movement sends out shock waves that are felt as an _____ at the surface.

    jerks   earthquake   towards   mantle

    At conservative plate boundaries, the plates are _____ past each other. This is not a smooth process and _____ builds up. When the pressure builds up over a long period of time, one _____ will suddenly jerk past the other. This causes _____ and an earthquake on the surface.

    pressure   plate   sliding   shock waves

National 5

1. A magnitude 8.2 earthquake occurred in the middle of the Pacific Ocean, while a magnitude 5.9 earthquake struck San Francisco. Which would be reported on the news in Britain – none, one or both? Give reasons for your answer.

2. Describe the relationship between plates, plate boundaries and earthquakes.

3. In your own words, describe each of the following terms:
   focus, epicentre, seismograph, Richter scale.

4. Describe the three types of shock waves.

5. Explain, in detail, why earthquakes occur at constructive plate boundaries.

6. At destructive plate boundaries, the crust moves in a series of jerks. Explain why it moves in this way.

7. Draw an annotated diagram to explain how earthquakes occur at conservative plate boundaries.

Activities

Activity A

The strongest earthquakes ever recorded on the Richter scale are shown in this table. Draw a bar graph in your notebook to show the strength of these earthquakes.

	Where?	Magnitude	When?
1	Chile	9.5	22 May 1960
2	Prince William Sound, Alaska	9.2	28 March 1964
3	Off the west coast of Northern Sumatra	9.1	26 December 2004
4	Near the east coast of Honshu, Japan	9.0	11 March 2011
5	Kamchatka, Far East Russia	9.0	4 November 1952
6	Off the coast of Maule, Chile	8.8	27 February 2010
7	Off the coast of Ecuador	8.8	31 January 1906
8	Rat Islands, Alaska	8.7	4 February 1965
9	Northern Sumatra, Indonesia	8.6	28 March 2005
10	Assam, Tibet	8.6	15 August 1950

Activity B

Using Figure 13.1 and an atlas, try to work out which two plates caused each of the earthquakes shown in the table above.

Now complete the 'I can do' boxes for this chapter.

Chapter 14

The cause of the Japan earthquake, 2011

This chapter looks at the 2011 Japan earthquake.

By the end of this chapter, you should be able to:

✓ explain why Japan experiences earthquakes
✓ describe the cause of the 2011 earthquake
✓ give reasons why predicting earthquakes is difficult.

The 2011 Japan earthquake

In 2011 Japan suffered its most powerful earthquake in a thousand years. The earthquake unleashed a tsunami up to 30 metres in height and resulted in the worst nuclear disaster in 25 years. This chapter explains why.

Japan is a small country in the Pacific Ocean in East Asia. It is made up of several islands with Honshu, Hokkaido, Kyushu and Shikoku being the four largest (see Figure 14.1).

Despite Japan's small land area, it has the tenth largest population in the world with over 127 million people. Nearly three-quarters of these people are squeezed into a narrow coastal strip because inland is very mountainous. Over 30 million people live in just one city, Tokyo, the largest urban area in the world.

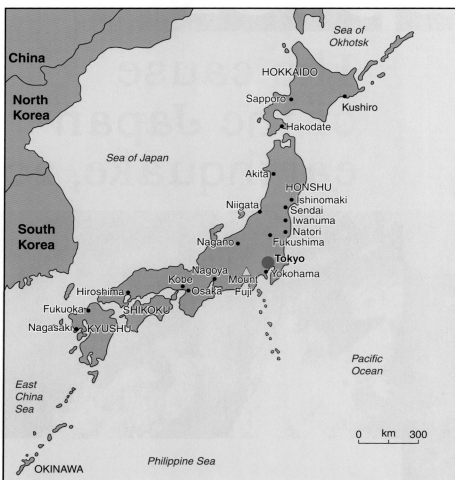

Figure 14.1
Japan

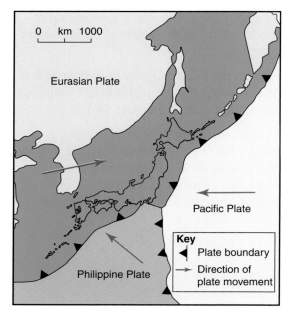

Figure 14.2
Crustal plates near Japan

Japan lies in an area where three crustal plates meet: the Eurasian Plate, Philippine Plate and Pacific Plate (see Figure 14.2). These are three destructive plate boundaries and, for this reason, Japan has over 1000 earthquakes each year. Some of these are only tremors; however, others are very strong, violent earthquakes. Table 14.1 shows the dates and magnitude of the ten largest earthquakes during the period 2010–2012.

When?	Magnitude
26 February 2010	7.0
21 December 2010	7.4
9 March 2011	7.2
11 March 2011 2.46p.m.	9.0
11 March 2011 3.08p.m.	7.4
11 March 2011 3.15p.m.	7.9
11 March 2011 3.25p.m.	7.4
7 April 2011	7.1
11 April 2011	7.1
10 July 2011	7.0
7 December 2012	7.3

Table 14.1
Dates and magnitude of ten largest earthquakes in Japan during 2010–2012

Did you know...?
A tsunami is a series of waves in the ocean but these are not ordinary waves caused by the wind. They are caused by an earthquake or volcanic eruption on the ocean bed. The tsunami waves are usually small in deep water but become much bigger in shallow water and can travel at great speeds.

Cause of the 2011 earthquake

At 2.46p.m. **on 11 March 2011, a magnitude 9.0 earthquake hit the north-east coast of Japan**; its epicentre was 70 km offshore and the focus was 6 km below the ocean bed. **The earthquake was caused by the Pacific Plate being pushed under the Eurasian Plate.** It moved 20–40 metres. The sudden jolting of the plates created the initial earthquake and was followed by more than 50 aftershocks greater than magnitude 6.0. The earthquake itself lasted for over five minutes and **the sudden uplift of the sea floor caused tsunami waves to spread out across the ocean at speeds of up to 800 km/h**. Tsunami warnings were issued as far away as Hawaii, Australia, Fiji, Mexico and Chile. The highest tsunami waves were over 30 metres high.

Did you know...?
Northeast Japan moved 2.4 metres nearer to the USA as a result of the earthquake.

Predicting the earthquake

Earthquakes are one of the most difficult natural disasters to predict but progress has been made in recent years. **Scientists had expected an earthquake of this magnitude in Japan for the following reasons:**

- Japan lies on the boundary of three crustal plates – all of which form destructive plate boundaries.

14

- They are also 'active' plate boundaries, with over 1000 earthquakes occurring there each year.
- In particular, in the previous week there had been several substantial earthquakes, one measuring 7.2 on the Richter scale.

Although people knew that a serious earthquake would happen in the near future, **experts could not predict exactly where and exactly when**. Scientists do not yet know when rocks that are under pressure at a plate boundary will suddenly move. For this earthquake, **seismometers detected strong movements in the crust one minute before the earthquake struck** which allowed warnings to be sent all over Japan immediately.

Although a big earthquake releases pressure at one part of the plate boundary, this pressure is then transferred to another part nearby, which increases the chances of another earthquake there in the near future.

National 4

1. Why does Japan experience so many earthquakes?
2. Look at Table 14.1 and rank the earthquakes in order of strength.
3. According to Table 14.1, which year had the most earthquakes? How many occurred?
4. Describe the reasons for the Japanese earthquake on 11 March 2011.
5. What is a tsunami?
6. What caused the tsunami on 11 March 2011?
7. Was it possible to predict the earthquake on 11 March 2011? Explain your answer.

National 5

1. Explain why Japan experiences so many severe earthquakes.
2. Using Table 14.1, rank the earthquakes in order of strength.
3. (a) Explain in detail why the Japanese earthquake on 11 March 2011 occurred.
   (b) On that day there were three more major earthquakes in that area. Why do you think that was?
4. Explain why a tsunami followed the earthquake.
5. Explain fully why it is so difficult to predict earthquakes.

Activities

Activity A

Carefully read the story below. You will notice that the story ends suddenly. You have to think of an appropriate ending for the story.

At 2.40p.m. on 11 March I was at school, sitting at my desk with all my other classmates. We were in geography learning about population distribution. I was listening carefully to my teacher as I really enjoy geography so I always pay attention. A few minutes later, out of nowhere, the ground started to shake, I could feel and hear my desk rattling and the windows in the classroom started to rattle too. Some of my classmates started to scream and we soon realised what was happening. My teacher tried to keep calm but I could tell by the expression on her face that she was panicking too. She shouted at us all to get under our desks. The walls were shaking and creaking and I could hear bits of the ceiling fall off and hit the floor and desks. The lights started to flicker and suddenly everything went quite dark …

Activity B

In each question below, there are two sentences describing two events connected with the earthquake. Copy the sentences carefully into your notebook and decide whether or not there is a connection between them. Next to each question write one of three letters:

- **M** if there *must* be a connection between the two.
- **C** if there *could* be a connection.
- **N** if there *cannot* be a connection between the two events and they just happened to occur at the same time.

1. Japan is made up of several small islands. Japan has a population of 127 million.
2. Japan lies near three destructive plate boundaries. Japan has many earthquakes and volcanoes.
3. Japan has many mountains and volcanoes. Japan has a very high population density.
4. Japan has many earthquakes. Japan has many tsunamis.
5. Japan experienced a very strong magnitude 9.0 earthquake. Japan lies near three destructive plate boundaries.
6. Japan gets over 1000 earthquakes each year. Predicting exactly when earthquakes will happen is impossible.

Now complete the 'I can do' boxes for this chapter.

Chapter 15

This chapter looks at the impact of the 2011 Japan earthquake.

The effects and management of the Japan earthquake, 20

By the end of this chapter, you should be able to:

✓ describe the impact of the earthquake on the landscape
✓ give examples of how the earthquake affected the people of Japan
✓ describe how Japan coped with this earthquake.

Impact on the landscape

The earthquake that hit Japan on 11 March 2011 released the same amount of energy as two million atomic bombs. The P waves travelled from the epicentre at 6 km/sec and the more destructive S waves followed at 3 km/sec. Within one minute of the earthquake, north-east Japan was shaking violently. **Buildings and bridges collapsed.** The shaking ground caused **landslides which buried cars, destroyed homes and blocked roads**. Electricity cables and gas pipes were badly damaged causing **thousands of homes and buildings to catch fire** and several oil refineries went up in flames. The earthquake was so violent that much of **the east coast of Japan dropped by one metre**.

The impact of the earthquake was very serious but worse was to come. The first tsunami waves hit the northeast coast 20 minutes after the earthquake. **The tsunami had waves of up to ten metres** crashing against the coastal towns. Tsunamis have devastating power because they are waves of debris not just water. Within the waves are cars, boats, whole houses and buildings and all their contents. **As the tsunami swept 10 km inland it demolished everything in its path – houses, farms and factories.**

In total, **2000 km of coastline was devastated** by the earthquake and the tsunami. In cities such as Iwanuma and Natori, homes, businesses and communications were completely obliterated by the waves (see Figures 14.1 and 15.4). **Several towns and villages were wiped off the map.** Altogether, at least **1.2 million buildings were damaged, destroyed, washed away or burnt down.**

Figure 15.1
Tsunami waves hitting Iwanuma, Japan

Figure 15.2
Tsunami waves hitting Natori, Japan

Meanwhile, the most serious effect of the earthquake was only just beginning. When the earthquake struck, the six reactors which make up Fukushima nuclear power plant automatically shut down but the cores of the reactors were still extremely hot and had to be cooled.

Did you know...?
The tsunami left 29 million cubic metres of waste – enough to fill Wembley stadium 25 times.

The earthquake caused power cuts which stopped the cooling systems from working. Emergency generators cooled them but these were then flooded by the tsunami. The reactors continued to heat up and reactors 1, 2 and 3 experienced full meltdown, which led to a number of chemical explosions. For many days it was feared that there would be a complete meltdown with large amounts of radioactive material being released into the atmosphere. However, sea water was finally used to cool down the reactors and save the plant, although it will never be used again.

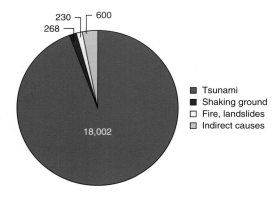

Figure 15.3
Numbers killed in 2011 Japan earthquake

Impact on the people

A few of the nuclear power plant's workers were severely injured or killed by the disaster. Some were exposed to high levels of radiation and **it has increased the risk of cancer in the local residents by about 1%**. Infants who were exposed to the radiation have the highest risk.

Altogether, **the earthquake and tsunami killed approximately 19,000 people** (see Figure 15.3). Ishinomaki city had the most recorded fatalities with 3735. People over age 65 made up 56% of the deaths.

The government closed down eleven other nuclear reactors for safety reasons, causing **power shortages** for several days. These affected the whole economy. In total **500,000 people were made homeless** by the earthquake and were in serious danger of hypothermia as temperatures plunged to −5 °C, with freezing winds, hail storms and thick snow affecting the country. People were forced to scavenge for food in the debris, as there were **major food shortages** across the affected areas. Four million people were without water and six million without electricity.

In total, the cost of the earthquake was as much as $300 billion, which makes it the costliest natural disaster ever and brought an economic crisis to Japan.

Planning for the earthquake

Japan is better prepared than any other country for earthquakes.

- Japan has an earthquake detection system and this picked up the earthquake as it happened. Automatic warnings were sent across the country within 31 seconds; one even interrupted Parliament.
- People in Japan know what to do in the event of an earthquake. Children have regular earthquake drills in school; offices, houses and schools all have emergency earthquake kits.
- There is a tsunami warning centre in Hawaii and a tsunami warning was sent to Japan 10 seconds after the earthquake. People were told to head for higher ground but in many areas this was too far away.
- There are tsunami drills so people know what to do.
- Coastal towns had already built sea walls up to 10 m high to keep out tsunami waves but, in some cases, these were not high enough because the earthquake had caused the land and the defence walls to sink.

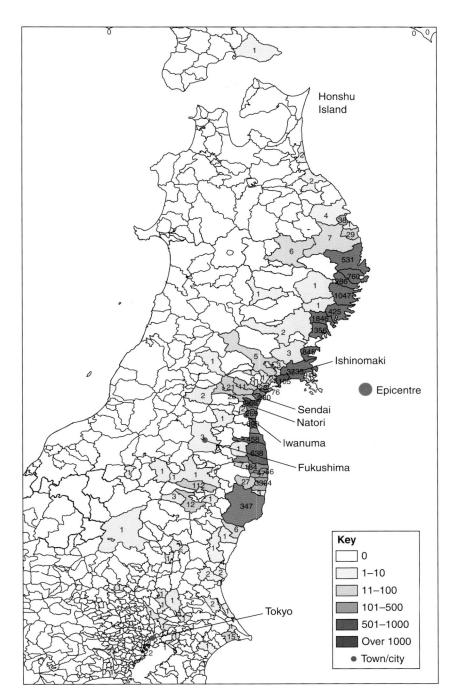

Figure 15.4
Numbers killed in 2011 Japan earthquake

■ Japan's planning measures saved many lives. Buildings are regularly checked to make sure they sway rather than shake during an earthquake and the majority of the buildings survived this earthquake because they were 'earthquake-proof'. The early warning earthquake and tsunami system worked. People knew of the earthquake immediately and they knew what to do when an earthquake occurs. Despite all these measures, 19,000 people died.

The relief effort

Immediately after the earthquake and tsunami struck, local people and charities started to provide assistance. The local authorities and the national government then stepped in and were quickly followed by promises of aid from other countries. Figure 15.5 shows examples of some of the different types of aid given by selected countries.

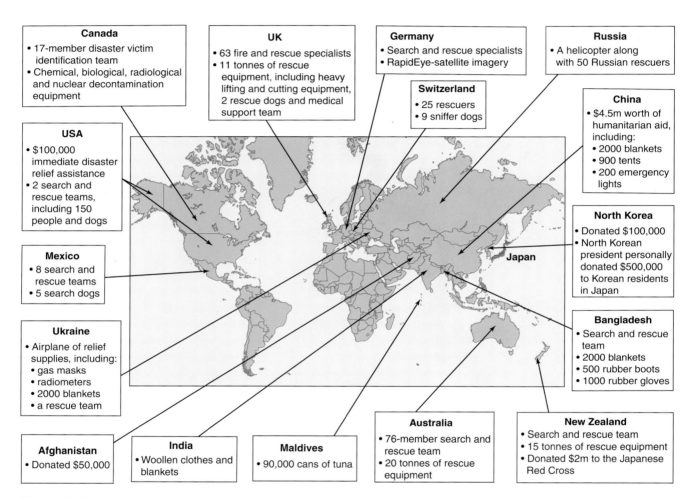

Canada
- 17-member disaster victim identification team
- Chemical, biological, radiological and nuclear decontamination equipment

UK
- 63 fire and rescue specialists
- 11 tonnes of rescue equipment, including heavy lifting and cutting equipment, 2 rescue dogs and medical support team

Germany
- Search and rescue specialists
- RapidEye-satellite imagery

Russia
- A helicopter along with 50 Russian rescuers

Switzerland
- 25 rescuers
- 9 sniffer dogs

China
- $4.5m worth of humanitarian aid, including:
  - 2000 blankets
  - 900 tents
  - 200 emergency lights

USA
- $100,000 immediate disaster relief assistance
- 2 search and rescue teams, including 150 people and dogs

North Korea
- Donated $100,000
- North Korean president personally donated $500,000 to Korean residents in Japan

Mexico
- 8 search and rescue teams
- 5 search dogs

Japan

Bangladesh
- Search and rescue team
- 2000 blankets
- 500 rubber boots
- 1000 rubber gloves

Ukraine
- Airplane of relief supplies, including:
  - gas masks
  - radiometers
  - 2000 blankets
  - a rescue team

Afghanistan
- Donated $50,000

India
- Woollen clothes and blankets

Maldives
- 90,000 cans of tuna

Australia
- 76-member search and rescue team
- 20 tonnes of rescue equipment

New Zealand
- Search and rescue team
- 15 tonnes of rescue equipment
- Donated $2m to the Japanese Red Cross

Figure 15.5
Examples of international aid given to Japan

National 4

Impact on the landscape	Impact on the people

1. Make a table in your notebook like the one above. Use the information in this chapter to complete your table.
2. Look at Figure 15.3. Which was more deadly – the earthquake or the tsunami? Explain why.
3. Why do you think so many older people died in the earthquake?
4. Japan was prepared for a severe earthquake. Describe two of the planning measures and explain how they saved lives.
5. Although Japan was well-prepared, many lives were lost. Explain why.
6. Using Figure 15.5, in your opinion what were the two most important types of aid that were given? Why?

National 5

1. Summarise the effect of the earthquake on the landscape.
2. Describe, in detail, the impact that the earthquake had on the people of Japan.
3. Look at Figure 15.3.
   (a) Describe what is shown.
   (b) Give reasons for the different causes of deaths shown.
4. Some people say that Japan's planning for this earthquake was very effective; others say it was not. Explain both points of view.
5. Using Figure 15.5, in your opinion what were the three most important types of aid that were given? Why?

Activities

Design and write a newspaper front page for the day after the Japan earthquake. You should:

- give your newspaper a title at the top and a date
- have an eye-catching headline underneath.

Now divide the rest of your page into two or three columns and have between two and four sections in each column. In each section write about one aspect of the earthquake, such as:

- where and when it happened
- its cause
- its strength
- the effects of the earthquake
- the tsunami
- the relief effort.

If possible, include a picture and a map.

Now complete the 'I can do' boxes for this chapter.

Chapter  16

This chapter looks at the location and features of tropical storms.

Tropical storms

By the end of this chapter, you should be able to:

- ✓ describe what a tropical storm is
- ✓ describe the main features and locations of tropical storms
- ✓ describe the conditions needed to create a tropical storm.

Tropical storms

Tropical storms are severe depressions in which **wind speeds reach over 60 km/h** but can often reach over 200 km/h. As Figure 16.1 shows, tropical storms are **found over oceans within 30 degrees of the equator**. They start on the eastern side of oceans and move westwards, before dying out over land. **When tropical storms reach 120 km/h, they are called hurricanes.** There are local names for hurricanes in different parts of the world, as shown in Figure 16.3 on page 83.

Main features of tropical storms

About 500 million people in 50 countries live in fear of tropical storms. They kill more people each year than earthquakes or volcanoes, yet some parts of a tropical storm are much more deadly than others. The main features of a tropical storm are described below and shown in Figure 16.1.

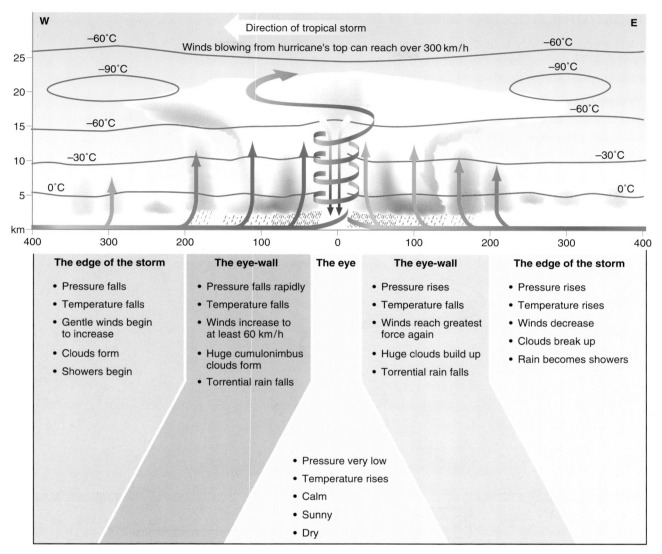

Figure 16.1
Main features of a tropical storm

1. As the storm approaches, the air pressure and temperature drop, while cloud cover and rainfall increase.
2. **Near the centre, at the eye-wall**, huge cumulonimbus clouds rise up, **torrential rain falls and wind speeds reach their maximum.**
3. **At the centre, the eye is calm, clear, warm and dry.**
4. After the centre is the other eye-wall and the same weather as in point 2 is experienced again, with towering clouds, very heavy rain and very strong winds.
5. At the edge of the storm, the air pressure and temperature rise, while cloud cover and rainfall decrease.

A tropical storm travels at about 10 km/h, but it can speed up or slow down quickly. The route a tropical storm takes is called a 'track' and **it can change direction suddenly**. On reaching coastal areas, **it can raise the level of the surface water** by up to ten metres. At high tides **this produces a storm surge**, which leads to severe flooding. Once a tropical storm reaches land it slows down, changes direction and quickly dies out. An average tropical storm lasts for one to two weeks.

Conditions needed for a tropical storm

Tropical storms are only found in certain areas of the world, as shown in Figure 16.3. These are the areas that have the necessary conditions for them to form. A tropical storm needs the following conditions to form:

1. **Warm seas, which have a surface temperature of 27 °C or more**, and warm water to a depth of at least 60 metres.
2. **Low air pressure**, with the air beginning to rise.
3. **Damp moist air** with a relative humidity of 60% or more.

Where these conditions are found, there are five stages in the formation of a tropical storm. These are shown in Figure 16.2.

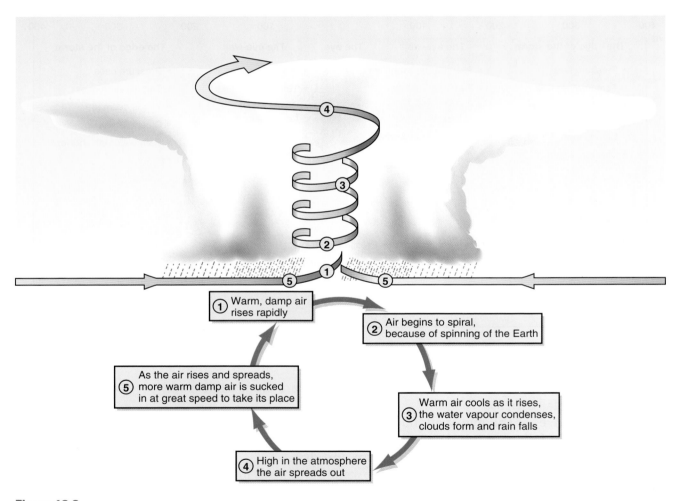

Figure 16.2
Stages in the formation of a tropical storm

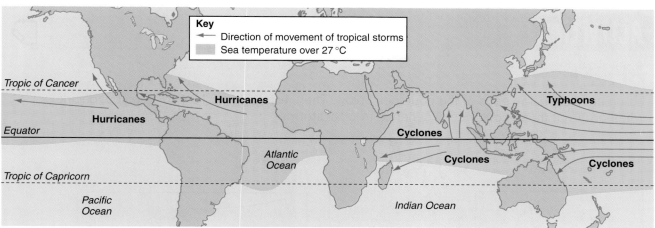

Figure 16.3
Distribution of tropical storms

National 4

1. What are tropical storms?
2. Describe the distribution of tropical storms in the world.
3. Describe the weather conditions at the eye-wall.
4. Where is the calmest weather found? Describe the weather experienced there.
5. What is a storm surge?
6. What are the three conditions needed for tropical storms to form?
7. What are the five stages in the formation of a tropical storm?

National 5

1. What is the difference between a tropical storm and a hurricane?
2. What name is given to a tropical storm in the
     (a) Atlantic Ocean
     (b) Indian Ocean
     (c) Pacific Ocean?
3. Which part of a tropical storm brings the worst weather? Describe, in detail, the weather it brings.
4. Where is the calmest weather found? Describe the weather experienced there.
5. Describe the movement of tropical storms.
6. Describe the conditions needed for tropical storms to form and explain why these conditions are needed.
7. Why does rapidly rising air lead to
     (a) heavy rain
     (b) very strong winds?

16

Activities

Activity A

Using an atlas, name each of the countries below. Then, using Figure 16.1 to help you, decide whether or not these countries experience tropical storms.

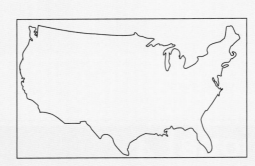

Activity B

Carefully read the statements below. You need to decide whether the statements are true or false. If the statements are false, you must correct the information to make them true.

1. Tropical storms are found within 40 degrees of the equator.
2. When tropical storms reach 120 km/h they are called hurricanes.
3. Wind speeds become stronger towards the centre of a storm.
4. Tropical storms travel at approximately 60 km/h.
5. Once tropical storms reach land they speed up.
6. Tropical storms need seas over 27 °C, warm water to a depth of 600 metres, high pressure and damp moist air.

Now complete the 'I can do' boxes for this chapter.

Chapter 17

This chapter looks at the causes of Hurricane Katrina.

The cause of Hurricane Katrina, 2005

By the end of this chapter, you should be able to:

✓ describe the differences between tropical storms and hurricanes
✓ explain the conditions that formed Hurricane Katrina
✓ describe the path of Hurricane Katrina.

Hurricane Katrina

On 29 August 2005 a Category 5 hurricane obliterated coastal states of the USA. The tropical storm was named Hurricane Katrina and it is recognised as one of the worst natural disasters in the history of the USA. A tropical storm becomes a hurricane once wind speeds reach 120 km/h.

Hurricane Katrina is the largest hurricane ever recorded to make landfall in the USA and it struck the states of Florida, Louisiana and Mississippi. The cause of Hurricane Katrina is shown in Figure 17.1. The timescale of Hurricane Katrina is described below. Figure 17.2 shows the path and strength of Hurricane Katrina.

July/August 2005

After a hot summer, the sea around the Bahamas reached 27 °C.

Water vapour evaporated from the sea, making the air above humid and damp.

23 August 2005

The air in contact with the sea became very hot and began to rise, creating low pressure.

Air was then sucked in over the sea to replace the rising air. A tropical depression had now formed. This was the birth of Hurricane Katrina.

As the humid air rose and cooled, the water vapour condensed, huge cumulonimbus clouds built up and torrential rain began to fall.

24 August 2005

By now, the hot air was rising even more rapidly, so air rushed in even faster, making stronger and stronger winds.

The wind speeds reached 60 km/h and Katrina became a tropical storm.

Katrina started to move west, towards the state of Florida.

25 August 2005

Wind speeds were increasing; when they reached 120 km/h Katrina became a Category 1 hurricane.

Katrina made landfall in southern Florida on 25 August as a Category 1 hurricane.

28 August 2005

By 28 August, wind speeds reached a peak of 266 km/h. Katrina was now a Category 5 hurricane.

29 August 2005

Katrina continued to head west, making landfall again on the 29 August.

Katrina had lost some energy at this point and was downgraded to a Category 3 hurricane when it hit Louisiana and Mississippi; however, wind speeds were still 200 km/h.

30 August 2005

Once over land, without any moist air, Katrina lost energy and began to slow down.

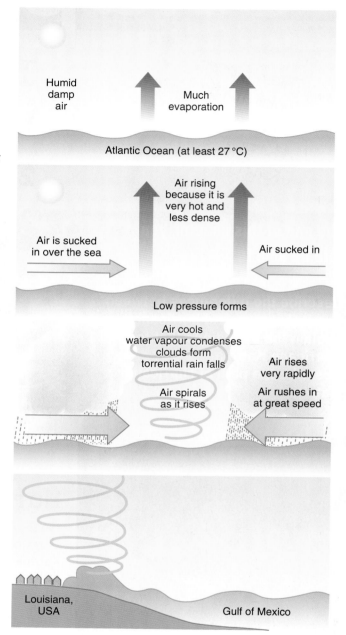

Figure 17.1
Formation of Hurricane Katrina

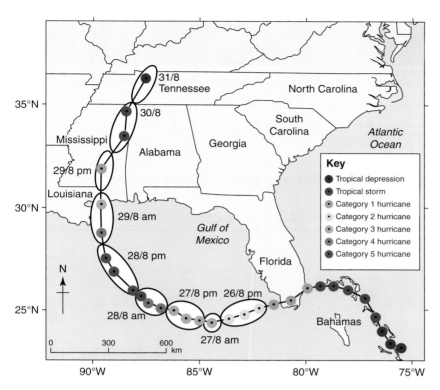

Figure 17.2
The path and strength of Hurricane Katrina

National 4

1. Which US states were affected by Hurricane Katrina?
2. Explain why air began to rise over the Bahamas during August 2005.
3. Explain how rising air led to the very strong winds of Hurricane Katrina.
4. At what wind speed did Katrina become
   (a) a tropical storm
   (b) a hurricane?
5. Using Figure 17.2, describe the path of Katrina.
6. Where did Katrina eventually lose energy on 30 August and why?
7. Copy Table 17.1 and complete it to show the movement of Katrina between 24 and 30 August 2005.

Date	Location of Katrina	Wind speed
24 August	Bahamas	60 km/h
25 August		
26 August		
27 August		
28 August		
29 August		
30 August		

Table 17.1

National 5

1. Describe the region affected by Hurricane Katrina, including the names of states, islands and seas.
2. Explain why Hurricane Katrina:
   (a) had such strong winds
   (b) brought heavy rain.
3. Using Figure 17.2, describe the path of Katrina.
4. Copy Table 17.1 and complete it to show the movement of Katrina between 24 and 30 August 2005.
5. Explain the changing wind speed of Katrina between 24 and 30 August.

Activities

Activity A

The statements below explain the formation of a hurricane. Rearrange the statements in the correct order.

- When wind speeds reach 60 km/h a tropical storm is formed.
- The air becomes very hot and starts to rise, creating low pressure.
- Evaporation occurs in seas where the temperature is over 27 °C.
- More and more air rushes in, which makes the winds become stronger and stronger.
- When wind speeds reach 120 km/h a hurricane forms.
- Air is sucked in under the rising air to create a tropical depression.

Activity B

Table 17.2 below shows the Saffir Simpson scale, which is used to measure the strength of a hurricane to produce a category numbered from 1–5. Using this, decide what category of hurricane produced the damage in pictures A, B and C. Explain your decision.

Category	Wind speeds (km/h)	Effects
1	120–155	Trees and power lines can be brought down. Damage to buildings is minimal. Flooding might occur.
2	155–180	Some structural damage to buildings. Widespread damage to trees and farmland. Power lines will be downed. Widespread flooding.
3	180–210	Significant damage to buildings. Mobile homes will be completely destroyed. Storm surges may cause extreme flooding.
4	210–250	Extensive damage to buildings. Major damage to area. Storm surges may cause extreme flooding.
5	More than 250	Complete devastation of buildings. Major roads destroyed or cut off. Vegetation is completely destroyed.

Table 17.2

17

Activities continued...

A

B

C

Now complete the 'I can do' boxes for this chapter.

Chapter  18

This chapter looks at the damage caused by Hurricane Katrina.

The impact and management of Hurricane Katrina, 2005

By the end of this chapter, you should be able to:

- ✓ explain the impact of Hurricane Katrina on the landscape
- ✓ give examples of the impact of Hurricane Katrina on the people
- ✓ describe the relief effort after Hurricane Katrina.

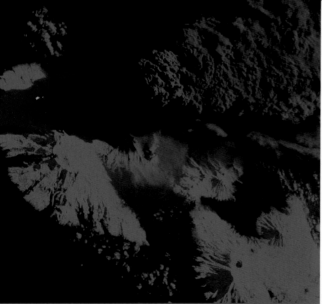

Impact on the landscape

On its journey across south USA, Hurricane Katrina lay waste about 230,000 km² but it reserved its worst damage for New Orleans, a city of over one million people, most of whom lived below sea level. The main states affected by Hurricane Katrina were southern Florida, Louisiana and Mississippi.

Southern Florida

Despite Katrina being only a Category 1 hurricane when it made landfall in Florida, the damage was still extensive. **The 130 km/h winds knocked down trees and power lines** and left 1.4 million people without electricity. The **torrential rain** – 350 mm in some places – **caused flooding in many areas and damage to homes and buildings**. An estimated $1.5 billion worth of damage was reported in Florida with most of the damage caused by the flooding.

Louisiana

When Katrina made landfall in Louisiana on 29 August, the winds were 200 km/h. This created an estimated **seven metre**

storm surge that flooded many coastal areas. **Eighty per cent of the city of New Orleans was flooded under six metres of water.** The widespread flooding resulted in the destruction of thousands of homes, buildings, roads and bridges. Power lines were also brought down, leaving 900,000 people without power. Katrina caused approximately $22 billion worth of damage in Louisiana.

Mississippi

A massive $125 billion worth of damage was reported in Mississippi. The 200 km/h winds and the **eight metre high storm surge destroyed bridges, houses, roads and buildings and washed boats, piers and cars up to six miles inland**. All 82 counties in Mississippi were declared disaster areas.

Figure 18.1
Widespread flooding caused by heavy rain and a seven metre storm surge left much of Louisiana underwater

Impact on the people

It is thought that Hurricane **Katrina killed 1836 people**, with most deaths occurring in Louisiana (1577). **Most people drowned** in the floodwater, although **collapsing buildings caused some deaths**. In Louisiana 33,544 people had to be rescued from rooftops or by boat as the floodwater meant they were completely isolated. Eight years later and 705 people are reported as 'still missing'.

In total, Hurricane Katrina affected over 15 million people. **Over 1 million people were made homeless. Hundreds of thousands of people were left unemployed and huge areas of farmland were completely destroyed.** Thirty offshore oil platforms were damaged by the hurricane and nine refineries were shut down. As a result, **oil production was reduced by 25% for six months.**

Did you know....?
Hurricanes used to be given girls' names because they were so unpredictable!

Planning for the hurricane

We all know how difficult it is for weather experts to forecast our weather. Hurricanes are even more difficult to predict than the weather systems which cross the British Isles, yet **a lot of sophisticated equipment is used** in the USA (see Figure 18.2).

Hundreds of **weather stations** on land and at sea record the weather as the hurricane approaches and passes over, giving information on its windspeed, wind direction, temperature and pressure	
Radiosonde balloons are sent into the hurricane carrying weather instruments and they send back information on temperature, pressure and humidity	
Radar is used to find out where the rain is falling and its intensity	
Satellites take photographs of the hurricane so that its speed and direction can be tracked	
Specially designed aircraft fly into hurricanes and record windspeed, wind direction and temperature	
Computers in the National Hurricane Center in Miami, USA process all these data and, based on how previous hurricanes have behaved (stored in their memory), they predict the hurricane's speed, strength and direction over the next few days	

Figure 18.2
Methods of forecasting hurricanes

Everyone in southern USA knows the likely time of year that hurricanes strike – commonly referred to as 'hurricane season' – because they understand what causes them. The government regularly informs people of how they should prepare for hurricanes; they even have a National Hurricane Preparedness Week before the hurricane season starts. But they know that **hurricanes are difficult to predict and plan for because they change direction and speed often**.

The USA has a **National Hurricane Center in Florida** which tracks all tropical storms. Special aircraft even fly into hurricanes to obtain weather data. The National Hurricane Center **accurately predicted the path and speed of Hurricane Katrina**.

Figure 18.3
Poster advertising National Hurricane Preparedness Week

The country's **National Weather Service was then able to inform people** of the hurricane's expected speed and route and the damage it was likely to cause. This saved millions of lives because plenty of warning and advice were given.

Because they were well-informed, the authorities were able to carry out a **compulsory evacuation of people from New Orleans**. This was planned in three phases, although there was a shortage of transport and fuel at the time of the evacuation. Nevertheless, 90% of the people left the city. For those unable to leave, the **Louisiana Superdome was used to house 26,000 people and provide them with food and water**.

Aid given following the hurricane

Hurricane Katrina is the costliest natural disaster, ever. President Bush and the US government were highly criticised for their response to Hurricane Katrina. **The American people did not think that the government responded quickly enough to the hurricane or with enough aid.**

As the world heard of the devastation caused by the hurricane, aid started pouring in from other countries, from charities such as the Red Cross and from international organisations such as the European Union and NATO.

> **Did you know...?**
> The US government refused help from some countries following Hurricane Katrina.

Different types of aid were needed. **First, hundreds of thousands of people in the affected states needed to be rescued.** The devastating winds and floodwaters had destroyed buildings and many people were trapped inside. Many countries, including the UK and China, sent rescue workers immediately. Others, such as Canada and Mexico, sent helicopters and other rescue equipment. Countries such as Australia and Afghanistan sent money.

Second, **aid was needed for the survivors**. With millions of people made homeless by the disaster, food, water, beds and blankets were urgently required. Many countries offered these immediately. Slovenia, for example, offered $120,000 worth of cots, mattresses, blankets, temporary shelters and first aid kits, while Pakistan sent doctors and paramedics. Without these many more people would have died.

The final type of aid needed was long-term aid so that the country could start rebuilding. Because so much was destroyed by the hurricane, everything needed to be rebuilt, including roads, schools, houses and hospitals. President Bush promised help to rebuild the affected states. However, **recovery was very slow**. The US government gave $110 billion to rebuild the affected areas and help the victims but, one year later, only $44 billion of this had actually been spent. Seven years after the hurricane, almost 5000 people remained homeless and 40,000 buildings were still abandoned.

National 4

1. Describe the damage to the landscape of southern Florida and Louisiana.
2. Which state had the most damage in terms of cost and how costly was it?
3. How many people died as a result of Hurricane Katrina and what caused the most deaths?
4. Choose two methods of forecasting hurricanes and describe how they do so.
5. What made planning for Hurricane Katrina difficult?
6. Describe the planning measures used in New Orleans.
7. Hurricane Katrina was well predicted. Describe what might have happened if the region had been completely unprepared for it.
8. Explain why short-term and long-term aid was needed after the hurricane.
9. Do you think that the aid effort was successful? Give reasons for your answer.

National 5

1. Compare the damage to the landscape of southern Florida and Louisiana and explain the differences.

2. Describe the effects of Hurricane Katrina on the people of southern USA.

3. Look at Figure 18.2. Which three methods of forecasting Hurricane Katrina do you think were the most useful? Explain your decision in detail.

4. The American people knew a severe hurricane was about to hit southern USA. Explain fully how this helped to reduce deaths and damage.

5. Describe the types of short-term aid given and explain in detail the reasons why they were needed.

6. Some countries sent money for long-term aid. Give examples of how this money should have been spent, and why.

7. How effective was the short-term and long-term aid effort?

Activities

Activity A

Use a map of the southern states of the USA, similar to the one here. Annotate the map to show the damage to each of the three states caused by Hurricane Katrina.

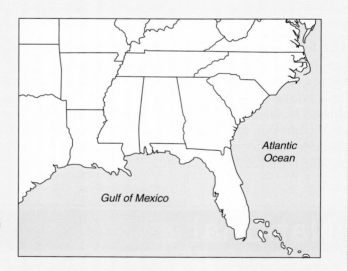

Atlantic Ocean

Gulf of Mexico

Annotate your map with the following information:

- Florida, Louisiana, Mississippi, Texas, Alabama
- Using Figure 17.2, draw on the path of Hurricane Katrina.
- Draw three boxes – one each for Florida, Louisiana and Mississippi – and inside each create a bulleted list of the damage caused by Katrina.

Activity B

Design a poster for a charity persuading people to help those affected by Hurricane Katrina.

Now complete the 'I can do' boxes for this chapter.

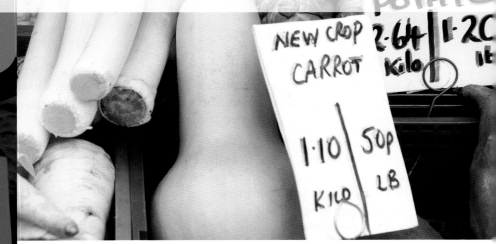

Chapter 19

This chapter looks at different types of employment across the world.

Global differences in employment

By the end of this chapter, you should be able to:

✓ give examples of primary, secondary and tertiary industry
✓ describe how the types of industry in a country change over time
✓ describe the countries which are important for the three types of industry.

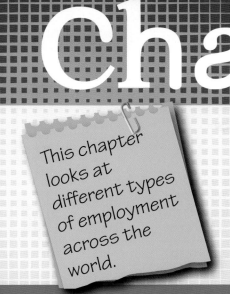

Types of industry

Hopefully when you leave education and start a career, there will be plenty of jobs from which to choose. Across the UK there are thousands of different types of jobs – from bank clerk to farmer to shop assistant to lawyer to international rock star. All of these jobs are types of industry and together they produce the country's wealth.

Because there are so many industries it is helpful to classify them into one of four types:

Primary industry

A primary industry produces primary goods. Primary goods have not been made but occur or grow naturally, for example coal, wheat and wood. Primary industries include farming, mining and forestry.

Secondary industry

A secondary industry produces secondary goods. Secondary goods have been made and are sometimes called manufactured goods. Primary goods provide the raw materials for making manufactured goods. For example, wheat is needed for making bread, wood for making paper and coal for making steel. Secondary industries include factories such as car assembly plants, steelworks and oil refineries.

Tertiary industry

A tertiary industry provides services. A service industry does not make or grow anything. Examples of services are health (doctor, nurse), education (teacher), transport (train driver, pilot) and entertainment (actor).

In recent years a new branch of service industry has emerged and has grown so quickly that it is now treated as a separate type of industry. This is referred to as a quaternary industry.

Quaternary industry

A quaternary industry provides quaternary services. Quaternary services provide expert information and advice. Examples of quaternary jobs are people who work in research and development, information technology and as consultants.

Some countries have more jobs in primary industry; others concentrate on secondary industry, while in many countries the majority of people work in tertiary industry. The proportion of people engaged in primary, secondary and tertiary jobs in a country is called its employment structure.

Changes in employment structure

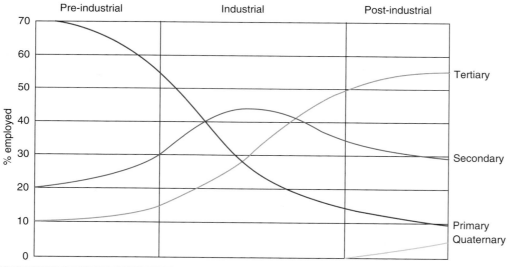

Figure 19.1
The Clark-Fisher model of employment change

The employment structure of a country (or area) changes over time. In the UK most people at one time were farmers, working in primary industry. When the Industrial Revolution took place 200 years ago, most people worked in factories, in secondary industry. Now most people's jobs involve providing a service, in tertiary industry. This is very similar to what has happened in other developed countries. Most countries have experienced a similar change in employment structure.

Figure 19.1 shows the Clark-Fisher model of employment change; it shows the usual or typical changes to a country's employment structure over time. It is made up of four graphs, one for each type of industry, and it shows how they change in importance over time (from left to right). Instead of putting in years, the graph has three time-periods – **pre-industrial**, before factories, followed by **industrial**, when there were many factories, and lastly **post-industrial**, when the old factories started to employ fewer people and close down.

Global differences in employment

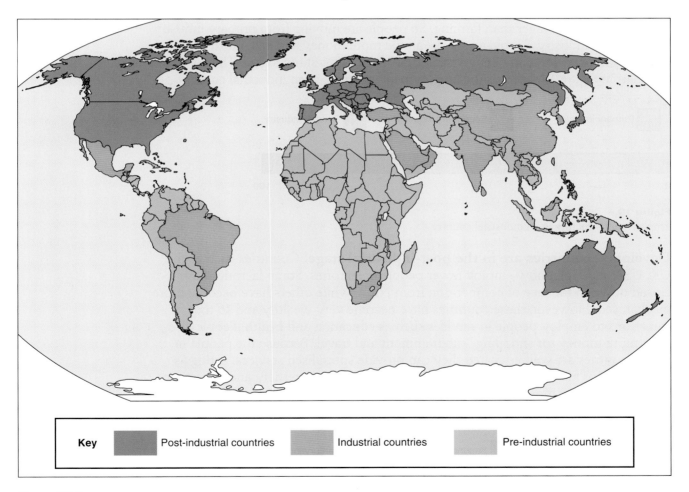

Key | Post-industrial countries | Industrial countries | Pre-industrial countries

Figure.19.2
Global differences in employment

Figure19.2 shows how employment differs around the world. **Some developing countries are still in the pre-industrial stage**, e.g. Malawi, Tanzania. They have many jobs in primary industry, especially farming. Farms are generally small with little mechanisation; there are few factories and offices in these countries. The divided bar chart in Figure 19.3 shows the employment structure in a pre-industrial country.

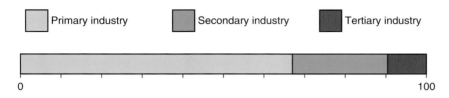

Figure 19.3
Employment structure in a pre-industrial country

Another group of developing countries, including Mexico and Brazil, are in the industrial stage. They are sometimes called the Newly Industrialised Countries (NICs). In recent years they have built factories in many of their cities which offer higher wages than farming. So people are moving from primary industry to secondary industry. People now have a little more money and can afford a few services (shops, entertainments) so the tertiary sector also starts to rise. The divided bar chart in Figure 19.4 shows the employment structure in an industrial country.

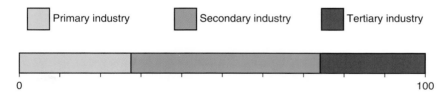

Figure 19.4
Employment structure in an industrial country

Developed countries are in the post-industrial stage. Countries such as the UK, USA and Japan now employ fewer people in factories. Some factories have closed down in the face of competition from NICs, while others have become more mechanised. However, these countries have become very wealthy and so they can afford to employ people in services such as education and health. People also have more money for shopping, entertainment and travel. Because the people in these countries are well-educated they can provide specialised services, acting as consultants or researchers or providing IT skills. As a result, tertiary and quaternary industries employ more people than secondary industry. The divided bar chart in Figure 19.5 shows the employment structure in a post-industrial country.

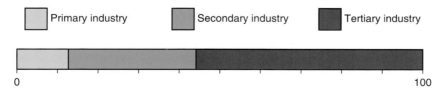

Figure 19.5
Employment structure in a post-industrial country

National 4

1. Give one example of a:
   (a) primary industry
   (b) secondary industry
   (c) tertiary industry.

2. What type of industry (primary, secondary or tertiary) do these people work in?
   (a) Travel agent
   (b) Fisherman
   (c) Lawyer
   (d) Whisky distiller

3. Look at Figure 19.1 and describe how the number of people in:
   (a) primary industry
   (b) tertiary industry
      changes over time.

4. Give one example of a country in the:
   (a) pre-industrial stage
   (b) industrial stage
   (c) post-industrial stage.

5. In which stage do most people work in secondary industry?

6. Why are there fewer people working in factories in developed countries?

7. Why do the number of jobs in services increase when people become wealthier?

National 5

1. What is the difference between:
   (a) a primary and a secondary industry
   (b) a tertiary and a quaternary industry?

2. What type of industry do these people work in?
   (a) Taxi driver
   (b) Fisherman
   (c) Research scientist
   (d) Forestry worker
   (e) Car mechanic

3. Look at Figure 19.1.
   (a) Describe the connection between the number of people in primary and tertiary industry over time.
   (b) Describe and explain the changes in the number of people in secondary industry over time.

4. Look at Figure 19.2. In which stages (pre-industrial, industrial or post-industrial) are the continents of:
   (a) Africa
   (b) Asia?

Activities

Activity A

Look at the table below and decide which stage each country is in – pre-industrial, industrial or post-industrial.

Country	Primary	Secondary	Tertiary	Quaternary	Stage (pre-industrial, industrial or post-industrial)
A	5%	30%	60%	5%	
B	80%	11%	9%	0%	
C	15%	45%	38%	2%	
D	30%	40%	30%	0%	
E	50%	20%	30%	0%	

Activity B

In which stage (pre-industrial, industrial or post-industrial) would you see these newspaper headlines?

1. The country is short of farm workers
2. Big increase in demand for housing in our cities
3. Record year for food imports
4. Derelict factories are an eye-sore in our cities
5. More and more research facilities being set up
6. Abandoned farms a common sight
7. Meeting for mineworkers needs a bigger venue
8. The country's first travel agent opens

Now complete the 'I can do' boxes for this chapter.

Chapter 20

This chapter looks at trade between countries.

World trade patterns

By the end of this chapter, you should be able to:

✓ explain the differences in the imports and exports of developed countries and developing countries
✓ describe the main barriers to world trade
✓ explain how world trade is unfair to developing countries.

International trade

All countries need to buy goods from other countries. What they buy is called their **imports**. To pay for these imports a country must sell goods abroad. These are called its **exports**. The movement of goods and services between countries is called **international trade**.

When a country's exports sell for more than the cost of their imports, they are said to have a **trade surplus**. If the exports are less than the cost of imports, they have a **trade deficit**. The difference between the cost of imports and the value of exports is called the **balance of trade**.

Global differences in trade

We found out in the last chapter that most people in developing countries work in primary industry, whereas people in developed countries work in secondary and tertiary industries. So **developing countries can export the primary goods** they produce but **they need to import manufactured goods** and services. **Developed countries**

can export some manufactured goods and services but **need to import primary goods** as well as other manufactured goods.

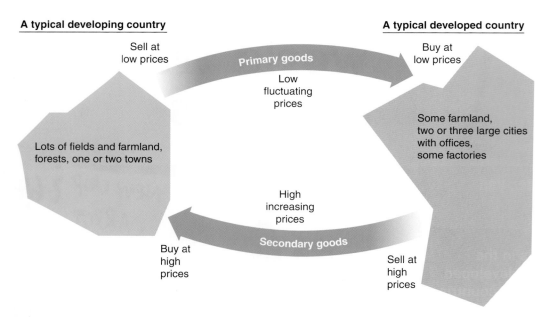

A typical developing country

Sell at low prices

Primary goods

Low fluctuating prices

Buy at low prices

Lots of fields and farmland, forests, one or two towns

A typical developed country

Some farmland, two or three large cities with offices, some factories

High increasing prices

Buy at high prices

Secondary goods

Sell at high prices

Figure 20.1
Trade between developing countries and developed countries

Developing countries fare badly in world trade. First, they have **fewer goods to export**, being poorer countries. Second, primary goods generally **fetch low prices** (see Figure 20.2). Third, these **prices go up and down** a lot so it is difficult to know how much money the country will make in a year.

Developed countries are in a much better position. They have **many goods to export,** manufactured goods and services usually **fetch high prices** (see Figure 20.2) and **prices are predictable** – they go up each year. (Just think of the manufactured goods your family buys.)

This means **developing countries can afford to buy few manufactured goods** and often have a trade deficit. Because they cannot afford to import much, they find it difficult to improve conditions in their country. If they decide to buy more than they sell, they go into debt and then have to repay these debts, usually with high interest. This gives them even less money in the future.

Developed countries can afford to buy many imports and often have a trade surplus. This makes it much easier for them to develop even further.

Sugar

A lot of sugar is traded around the world. As with most primary products, developing countries export more sugar than developed countries but developed countries import more sugar than developing countries (see Figure 20.2). The price of sugar fluctuates; for example in 2000 the price was only half that in 1998. In 2006 the price nearly doubled but by 2007 it was the same price as in 1998. Companies in developed countries make packets of sugar (sugar cubes, granulated sugar) from raw sugar and sell them in supermarkets. This is a secondary product. The price of a packet of sugar has gone up every year between 1998 and 2007.

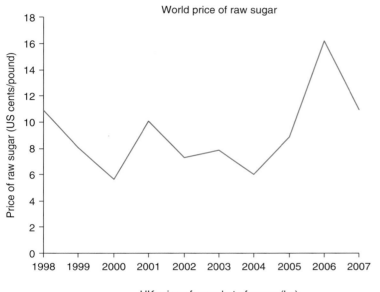

Main sugar exporters
Brazil
Thailand
Australia
India
Mexico

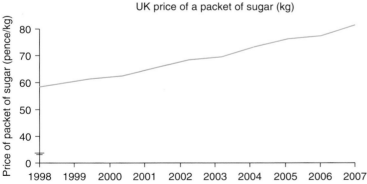

Main sugar importers
European Union
USA
Russia
Indonesia
China

Figure 20.2
Price of raw sugar and packets of sugar (1998–2007)

Free trade and trade barriers

Free trade occurs when goods can be exported to another country without any restrictions. With free trade a country can export all its goods and this will bring it more money. But it also has to accept imports from every country and it may not want to do this. This is because imported goods will mean increased competition for goods made in that country. So those companies will sell fewer goods, make less profit and employ fewer people.

Many countries try to protect their companies from imports in three ways: they can introduce trade barriers (quotas and tariffs) and subsidies.

Quotas

A quota is a limit placed on the amount of goods being imported from another country. For example, the USA has a quota on how many socks it imports from China. This makes it easier for American companies to sell socks in the USA.

Tariffs

A tariff is a tax or charge placed on goods imported from another country. For example, in 2010 China imposed a tariff on chicken products from the USA of over 100%. This more than doubled the price, which meant far fewer chicken products were sold. This made it easier for China's companies to sell their chicken products.

Did you know...?
The Queen received £7 million over ten years in farming subsidies funded by taxpayers. The Duke of Westminster – one of Britain's richest men – has been given around £6 million in farming subsidies.

Subsidies

A subsidy is a grant of money given to a company (usually farmers) for growing or producing a certain product, such as wheat, cotton, cattle or sheep. For example, cotton farmers in the European Union (EU) were given subsidies of £400 million in 2010. This has three effects:

1. Our farmers produce more of these products (e.g. wheat, cotton) and can sell them cheaply, which makes it difficult for other countries to sell their products here.
2. Increasing the amount produced lowers the price and makes it difficult for other countries to make a profit.
3. Our farmers can now sell these products at low prices in other countries, which often undercuts local farmers' prices. This is called dumping.

When countries use these methods to reduce trade and benefit their companies, they are called protectionist policies.

Did you know...?
The average subsidy given to every farm in the EU in 2010 was over £10,000.

National 4

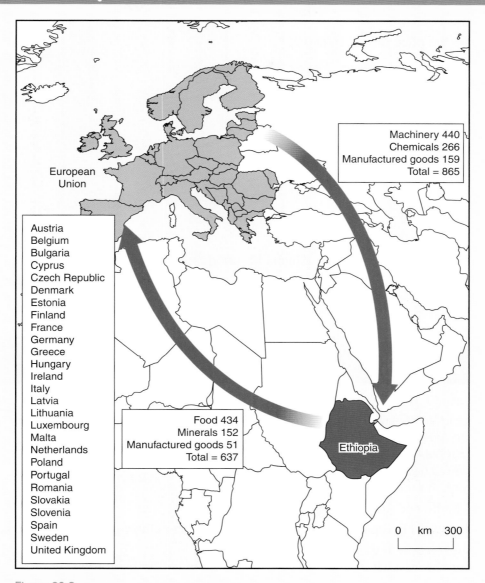

European
Union

Machinery 440
Chemicals 266
Manufactured goods 159
Total = 865

Austria
Belgium
Bulgaria
Cyprus
Czech Republic
Denmark
Estonia
Finland
France
Germany
Greece
Hungary
Ireland
Italy
Latvia
Lithuania
Luxembourg
Malta
Netherlands
Poland
Portugal
Romania
Slovakia
Slovenia
Spain
Sweden
United Kingdom

Food 434
Minerals 152
Manufactured goods 51
Total = 637

Ethiopia

0   km   300

Figure 20.3
Trade between the EU and Ethiopia in € millions (2011)

1. Look at Figure 20.3.
   (a) What are Ethiopia's main exports to the EU?
   (b) A typical developing country exports primary goods. Is this true for Ethiopia? Give reasons for your answer.
   (c) What are Ethiopia's main imports from the EU?
   (d) A typical developing country imports secondary goods. Is this true for Ethiopia? Give reasons for your answer.
2. What is a trade surplus and a trade deficit?
3. Does Ethiopia have a trade surplus with the European Union? Explain your answer.

National 4 continued...

4. Look at Figure 20.2.
   (a) What was the price of raw sugar in:
      (i)   1998
      (ii) 2000
      (iii) 2004
      (iv) 2006?
   (b) What was the price of a packet of sugar in:
      (i)   1998
      (ii) 2000
      (iii) 2004
      (iv) 2006?
5. Why is it better to sell packets of sugar than raw sugar?
6. Name one type of trade barrier.
7. Explain how a trade barrier makes it difficult for goods to be imported.
8. A subsidy is money given to farmers for growing a crop. Why does this make it difficult for other countries to sell that crop?

National 5

1. Look at Figure 20.3.
   (a) Describe Ethiopia's main imports from and exports to the EU in terms of primary and secondary goods.
   (b) Does Ethiopia have a trade surplus with the EU?
   (c) Does the EU have a trade surplus with Ethiopia?
      Give a reason for each answer.
2. Explain why tariffs and quotas are called trade barriers.
3. 'The average cow in Europe earns more than most people in the world – £2.20 per day in 2008.'
   (a) Explain how a cow in the EU 'earns' money.
   (b) Why does this make it difficult for other countries to export meat to EU countries?

Activities

Activity A

Between 2001 and 2004 the price of raw sugar was reduced to almost half its original price. Consider the knock-on effects of this. What would be the effect on the following people?

- Farmers in Central America growing raw sugar.
- Farmworkers in Central America growing raw sugar.
- Owners of the factory in the UK making packets of sugar (see Figure 20.2).
- Workers in the factory in the UK making packets of sugar.
- People in the UK buying packets of sugar.

Activities continued...

Activity B

1. Raw sugar is extremely important to the South American country of Guyana because it makes up nearly half its exports. How is Guyana affected when the price of sugar drops sharply?
2. The price of packets of sugar nearly always rises. Why doesn't Guyana make and export packets of sugar instead? Try and think of several points to make.

Activity C

1. The dates and events shown in the table below have been jumbled up. Match the correct date to each event.
2. When you have put the events in the correct order, write them down as a paragraph and make up a suitable heading.

Date	Event
1950	Japan's car makers find it easier to sell cars.
1950s	Japan is the world's biggest car maker.
1960s	USA is the world's biggest car maker.
1970s	Japan replaces quotas with tariffs on imported cars.
1980	Japan imposes quotas on imported cars.

Activity D

The sketch below shows a customs official talking to someone wanting to import goods into the country. The country has tariffs for those goods. Write down what the customs official might say if there were:

(a) quotas
(b) subsidies
instead of tariffs.

'They cost £10, do they? Well you need to pay us £3 and sell them for £13. Ours cost £12, by the way.'

Now complete the 'I can do' boxes for this chapter.

Chapter 21

This chapter looks at the recent rapid increase in world trade.

The growth in world trade

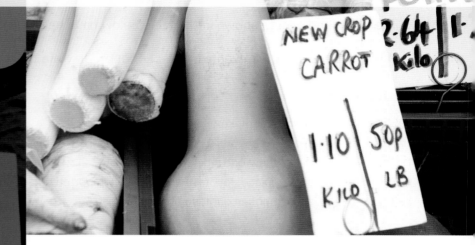

By the end of this chapter, you should be able to:

- ✓ explain why countries trade more now
- ✓ understand the meaning of countries being *interdependent*
- ✓ understand the meaning of *globalisation*.

World trade

Trade is good. It means that a country sells things it is good at producing so it can buy things it doesn't produce. Therefore people have more goods altogether.

When a country trades with other countries they become interdependent – they need each other.

The UK has always traded with other countries but recently this level of trade has increased. The same is true for almost every country in the world. This increase in world trade is shown in Figure 21.1. It shows that world trade doubled in the twelve years between 1990 and 2002 and doubled again in the next five years, with the biggest increase occurring between 2010 and 2011.

In the UK we now trade more and with more countries because:

1. **More countries are making goods** and providing services than ever before. For example, in 1950 most cars in the UK were made in the UK. We were the second biggest car manufacturer in the world (although altogether there were only five major car manufacturing countries). In 2011 we were 14th out of 37 major car manufacturing countries.

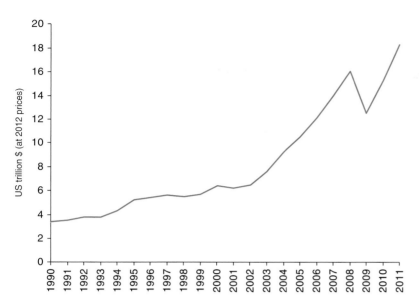

Figure 21.1
Changes in the amount of world trade

2. **We can communicate with people in countries much faster**, more reliably and much more cheaply using fax, the internet, email and mobile phones. For example, in 1980 if you wanted to buy goods from a company in Asia, a 10-minute phone call from the UK to Asia would cost you £20; a letter might take two weeks to reach there. Now an email gives the same information in seconds at a fraction of the price.
3. **It is quicker for people to travel** and for goods to be sent between countries. There are more flights than ever before, more lorries, more trains and more ships and they all go faster than ever before.
4. **It is relatively cheaper for people to travel** and for goods to be transported. Ships and lorries have become larger, making it cheaper to transport goods. There are now low-cost flights and cheaper cars.
5. **There is more free trade** and fewer tariffs and quotas reducing imports to countries.

eBay™ is an example of a modern company which makes it easy and quick for all of us to trade goods with people throughout the world. It started in 1995 and in 2012 eBay's net revenue was $14 billion.

eBay Mark is a trademark of eBay Inc.

Globalisation

It is not just goods that we trade more with other countries. There is also trade in services, for example Manchester United playing a friendly football game in China, a world tour by a British rock group, people using their IT skills in another country. So ideas and information spread much more quickly between countries and, of course, people travel more for work or leisure. This means more money is also being sent around the world. We are all becoming much more interdependent. We need and use other countries more, and other countries need us.

Figure 21.2

This recent big increase in trade has led to globalisation. **Globalisation is the name given to the way in which people and countries are increasingly connected and interdependent.** We are much more connected with other countries than our parents or grandparents were.

Return service

Fifty years ago we made tennis balls in the UK. Now the wool is brought 18,000 km from New Zealand, made into felt in the UK, sent back 11,000 km to the Philippines to be glued to the rubber centre (which comes from Malaysia), made into tennis balls and then sent back to the UK to be sold.

National 4

1. Look at Figure 21.1. What was the value of world trade in
   (a) 1991
   (b) 2001
   (c) 2011?
2. What is meant by globalisation?
3. What is meant by countries being interdependent?
4. More countries in the world make goods now. Explain how this increases world trade.
5. Look at the information on making a tennis ball.
   (a) Explain why making a tennis ball involves so many countries. In your answer you should mention how easy it is to send information now and how cheap it is to send goods long distance.
   (b) The tennis balls are still British-owned. The company's managers will need to visit their operations abroad. Is this easier to do now than 50 years ago? Explain your answer.
6. Give two examples of services that can be sold abroad.

National 5

1. Describe the changes in world trade shown in Figure 21.1.
2. What is meant by globalisation?
3. In 1950 most cars in the UK were made in the UK. Now most of the cars we buy are made abroad. Suggest reasons why.
4. Look at the information on making a tennis ball.
   (a) Explain how the UK and the Philippines are interdependent.
   (b) Explain why 50 years ago it would not have been possible to use so many countries when making tennis balls.
5. Give other examples of international trade now, apart from manufactured goods.

Activities

Making a pair of denim jeans

The jeans are made in Tunisia. To make them, the factory needs to import denim from a factory in Italy. The factory in Italy has made the denim, using:

- cotton from Benin
- softer cotton from Pakistan
- cotton thread from Hungary
- pumice from Turkey, to stonewash the jeans
- indigo from Germany, to dye the cotton to make denim
- other dyes from Spain
- zips from Japan
- brass rivets from Germany (the brass is made from copper which comes from Namibia).

Once the jeans are made they are packaged, using packaging from China. The jeans are then sent to a warehouse in France to be stored. Finally, the jeans are sent to the UK to be sold.

1. Look at the steps involved in making a pair of jeans. Draw a diagram or map with arrows (a flow map) to show all the countries involved in the making and selling of the jeans.
2. Give your diagram a title which includes the word globalisation.
3. Do you think it is good that so many countries are involved?

Now complete the 'I can do' boxes for this chapter.

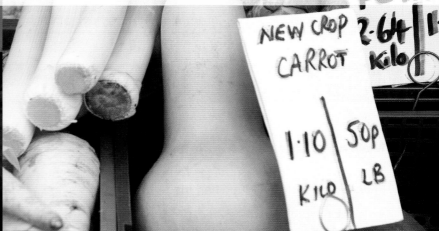

Chapter 22

Globalisation

This chapter looks at the benefits and drawbacks of globalisation.

By the end of this chapter, you should be able to:

✓ describe how globalisation benefits people in rich and poor countries
✓ describe the problems globalisation brings to people and countries
✓ explain the effects of globalisation on the environment.

Benefits of globalisation

We began Chapter 21 by saying that trade is good; in which case, globalisation must be even better – potentially. Here are some of the benefits associated with globalisation:

- We have the opportunity now to **buy many more products** and mostly **more cheaply** than ever before.
- We **know more about other countries** and we **visit them more often**.
- It has given **more employment opportunities abroad** for British people. We can eat at Japanese, Thai, Indian and Chinese restaurants without going any further than the High Street.
- From a world point of view, globalisation means **developing countries are able to trade more** and people can migrate to other countries to find work. There are now 35 million international migrant workers in the world. This should help to improve standards of living in developing countries.
- As a result of globalisation, large companies have set up factories and operations in developing countries and created **more employment in those countries**.

■ As companies become bigger and bigger, they can afford to spend **more money on research** and innovation, for example disease cures and green technology. These inventions quickly spread to and **benefit developing countries**, thanks to globalisation.

But, there are problems with globalisation.

Did you know....?
The top three companies in the world produce more in value in a year than the whole of Africa.

Problems associated with globalisation

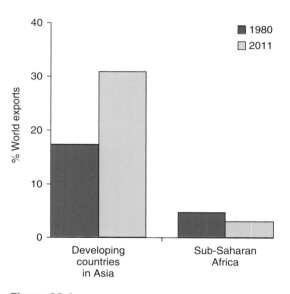

Figure 22.1
Trends in world exports

Figure 22.2
Trends in world income

Such a huge change inevitably causes problems. These are the main arguments against globalisation:

■ **It has made differences between countries greater.** Figure 22.1 and Figure 22.2 show that developing countries in Asia (e.g. China, South Korea, Singapore and India) have benefited from globalisation. They have nearly doubled their share of world trade and world income since 1980. Meanwhile the poorest countries have not had the skills and technology to benefit. For example, countries in sub-Saharan Africa have seen their share of world trade and world income decline. The number of extremely poor people in those countries has doubled since 1980.

■ Because they have more countries with which to trade, some **companies have become bigger and bigger**. Very large companies can produce their goods more cheaply; this means they can undercut smaller companies and drive them out of business. **This reduces competition and choice.** The effects of this can be seen in every high street and shopping mall where the same chain stores can be found. Figure 22.3 shows some of the shops in a shopping mall in Mumbai. It is the same in Shanghai, Rio de Janeiro and

Sydney. Over the last 20 years, not just in shopping but in other areas such as soft drinks, footwear and supermarkets, some companies have become giants and now dominate the others. Most of these are transnational companies.

- **The biggest companies are now found all over the world.** You will see people drinking Coca-Cola, eating McDonald's burgers and wearing Nike T-shirts in the Sahara desert, the Himalaya mountains and the Amazon rainforest, as well as the streets of New York. This can lead to **a loss of local cultures, as they are replaced by the same global culture** – same foods, same clothes, same shoes, same coffee shops everywhere. It is sometimes called the '"McDonaldisation" of the world'.

Figure 22.3
Shops in the InOrbit shopping mall in Mumbai, for example, include Pizza Hut, Benetton, Marks & Spencer, Costa Coffee, The Body Shop, KFC, Reebok and Accessorize. This photo was taken in the Manar Mall, Ras al-Khaimah, part of the United Arab Emirates.

- **It has badly affected the environment:**
  - More trade is taking place, so **minerals are being used up more quickly**. At the present rate, oil will only last until 2056, natural gas until 2070 and phosphorus soon after, followed by coal.
  - **Forests are being cut down** to make way for mines or roads or just to trade the timber. Every year 16 million hectares of forest are cut down.
  - Underground **water reserves are being pumped dry**. By 2025 it is thought that nearly 2 billion people will be living where there is not enough water.
  - **More greenhouse gases are being released into the atmosphere** because more fossil fuels are being used to transport goods around the world and there are more factories burning fossil fuels. Developed countries are responsible for 60% of the greenhouse gases going into the atmosphere.
  - **There are fewer habitats for wildlife** because of the increase in pollution and the loss of woodland and wetlands; this is affecting over 80 per cent of all threatened birds, mammals and plants.
  - **There has been over-fishing** with nine of the 17 major fishing grounds in the world now in decline.
  - **Fertile soils are being ruined** because they are being farmed too intensively.
  - **Deserts are spreading** as soil is ruined and vegetation cut down. The Sahara is growing at a rate of over 10 kilometres per year.

Rapid industrialisation in Mexico City

Mexico City has seen a rapid growth in factories which has attracted over 1000 people per day to the city. The factories, lorries and cars cause pollution which is trapped by the surrounding mountains turning it into a dangerous smog. The factories and the 20 million people here have to take increasing amounts of water from underground, causing the land to sink. Once they have used it, they dump it full of pollutants into the rivers. The city now produces 11,000 tonnes of waste every day but collects only 10,000 tonnes per day.

National 4

1. Describe three benefits of globalisation to people in the UK.
2. Describe three benefits of globalisation to people in developing countries.
3. Describe what is shown in Figure 22.2.
4. Explain how big companies make it difficult for smaller companies to make a profit.
5. Look at the information box on Mexico City. Mexico City has seen rapid growth in industry in recent years.
   (a) What effects has this had on Mexico City's environment?
   (b) Describe two other effects on the Mexican landscape which have resulted from globalisation.

National 5

1. In what ways has globalisation improved the lives of people in the UK?
2. Some people argue that globalisation has been good for developing countries. Give reasons why they say this.
3. Describe what is shown in Figure 22.1 and Figure 22.2.
4. In recent years some companies have become huge. Describe some disadvantages this has brought.
5. Look at the information box on Mexico City.
   (a) List the environmental problems that industry has brought to Mexico City.
   (b) Which of these problems are typical of globalisation?

Activities

Activity A

Figure 22.4
A town planning meeting

Figure 22.4 shows a planning meeting in a town in the UK. Starbucks has applied to take over an empty shop in the High Street. Give the different views that people around the table might have about this application. Write their short comments within speech bubbles. (One person's comment can be answering someone else's statement.)

Activities continued...

Activity B

Complete this sentence in two contrasting ways – one positive and one negative:

'If globalisation continues at this rate for the next 20 years …'

Now complete the 'I can do' boxes for this chapter.

Chapter 23

Transnational corporations

This chapter looks at transnational corporations.

By the end of this chapter, you should be able to:

✓ give a definition and examples of transnational corporations (TNCs)
✓ describe the benefits TNCs bring to countries
✓ describe the main criticisms of TNCs.

Did you know...?
Of the top 100 economies in the world, 51 are companies. The other 49 are countries.

Transnational corporations

Globalisation has led to big companies becoming even bigger. The biggest companies in the world are **transnational corporations (TNCs)**, similar to multinational companies (MNCs). They **have operations and do business in several countries**. Examples include Shell, BP, Nissan, Volkswagen, Sony and IBM. There were 7000 TNCs in 1970 but, helped by globalisation, this number had increased to nearly 80,000 in 2012. Of the 200 biggest TNCs, 199 have their headquarters in a developed country. And the top 500 TNCs are responsible for 70% of the world's trade.

Companies set up operations in other countries for several reasons:

■ It may be nearer to their raw materials and so saves transport costs.
■ Labour is cheaper there.
■ Tax rates may be lower.
■ It enables them to sell their product in those countries.

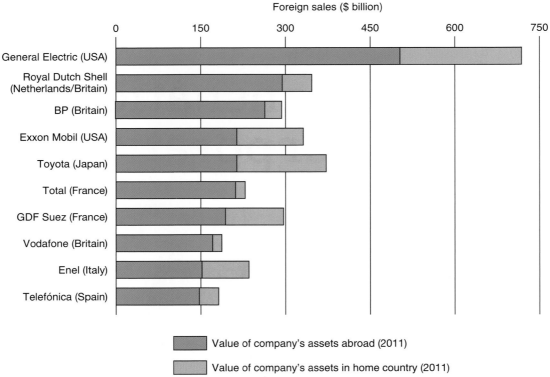

Figure 23.1
Top ten transnational corporations (2011)

Benefits of transnational companies

TNCs can bring many benefits to the countries where they set up:

- They can bring many **direct jobs** (e.g. working for the company) **and indirect jobs** (e.g. building the factory, bringing in goods by lorry, train or ship, creating more business for local suppliers).
- TNCs have to **pay taxes** and this gives the government of a country more money to spend on its people.
- TNCs may need to **train their workers** and so they bring new skills to the country.
- TNCs often **improve transport links** in the country.

Criticisms of transnational companies

- Although they increase employment, TNCs often bring in their own people to do the skilled, better-paid jobs.
- Profits go to shareholders in the country where the company is based – a developed country.
- TNCs very quickly close down their factories or mines in a country when they become unprofitable. This can cause sudden and very high unemployment, which increases poverty in that area of the country. This can happen in developed countries as well as developing countries, although in a developing country the closure of a big factory can affect the whole country, not just the immediate area near the factory.

Figure 23.2
Gold mining in Peru is carried out by TNCs, which employ 5000 people
directly and 8000 indirectly, but it causes huge environmental damage

- TNCs can be so important to a developing country that it is possible for them
  to influence the government of that country, making sure it makes decisions
  which favour the TNC, for example by making environmental laws less strict
  and imposing lower taxes for TNCs.
- In some cases the working conditions of the employees can be very poor.
- Like all large companies, TNCs sometimes drive smaller companies out of
  business (see Chapter 22).

National 4

1. Explain what is meant by a transnational corporation and give two examples.
2. Why are TNCs so important?
3. Give two reasons why transnational companies set up factories in other countries.
4. What are the two biggest benefits of a transnational corporation opening a factory in a developing
   country? Give reasons for your answers.
5. What are the two most serious problems associated with TNCs? Give reasons for your answers.

National 5

1. Transnational corporations are very big companies.
   (a) What is their exact definition?
   (b) Give evidence that shows how big they are.
2. Why have so many transnational companies moved their factories from a developed country to a
   developing country in recent years?
3. What is the difference between direct and indirect jobs?
4. A transnational corporation in a developing country is making large profits and increasing its
   workforce, but sometimes the developing country benefits only a little from this. Why?

Activities

Activity A

You are the president of a developing country. A transnational company wishes to set up a large mine in your country. It says it will employ up to 1000 workers and will build a new road and improve facilities at the nearest port. However, the TNC is concerned that your country's tax on companies (corporation tax) is too high. It says that if you would just reduce your corporation tax from 30% to 10%, it would start building its factory immediately.

    (a) Would you agree to reduce corporation tax?

    (b) Explain your reasons why you would or would not agree to this request. Consider all the possibilities.

Activity B

1. An American clothing company opens a factory making T-shirts in a developing country. Workers are paid £2 per day, on average, and have to work at least 12 hours a day. The factory has been built very cheaply, conditions inside are very hot and it can be dangerous. There have been many accidents and injuries. Because the factory in the developing country is profitable, the company has closed down its factory in Scotland.

    (a) Who benefits from this company's move?

    (b) Who loses out?

2. The clothing factory is in a town near the coast. The town is growing rapidly as people move there from nearby farming areas where they are only paid £1 per day, on average.

    (a) Does this make any difference to your answer for question 1(a)?

    (b) Explain why it does or does not make any difference to your answer for question 1(a).

Now complete the 'I can do' boxes for this chapter.

Chapter 24

Coca-Cola – a transnational corporation

This chapter looks at Coca-Cola, a transnational corporation.

By the end of this chapter, you should be able to:

✓ give evidence that Coca-Cola is a TNC
✓ describe the benefits that Coca-Cola, as a TNC, has brought to other countries
✓ describe some of the problems Coca-Cola has created in other countries.

Did you know...?
Coca-Cola is sold in every country in the world except two – Cuba and North Korea.

Coca-Cola is the number one manufacturer of soft drinks in the world. It is the world's best known brand. Its logo is recognised by 96% of all the people in the world.

Date	Event
1886	First Coca-Cola sold; initially marketed as a medicine for five cents a glass
1906	Opened its first three factories abroad (Canada, Cuba, Panama)
1912	First factory opened in Asia – the Philippines
1919	Opened a factory in Europe (France), followed by nine more countries
1930	Factories in 27 countries
1940	Factories in 44 countries
1950s	Began to open factories in Africa; now has factories in 50 countries
2013	Research centres opened in Turkey and China
2013	There are 275 factories in approximately 85 countries around the world

Table 24.1
Growth of Coca-Cola as a transnational company

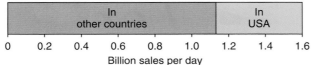

Figure 24.1
Sales of Coca-Cola per day

Coca-Cola ingredients

- Kola nuts (rich in caffeine)
- Extracts from coca leaves
- Sugar
- The rest is secret!

Figure 24.2
The iconic Coca-Cola bottle

Figure 24.3
Giant Coca-Cola can in the Atacama desert

Coca-Cola facts

- 92,800 employees worldwide
- It made profits of $18 billion in 2001 and $47 billion in 2011

Did you know...?
Coca-Cola is the biggest employer in Africa.

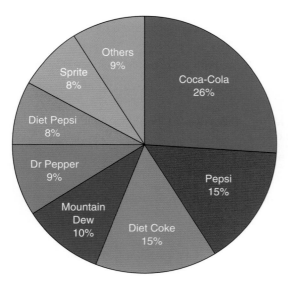

Figure 24.4
Most popular soft drinks in the world

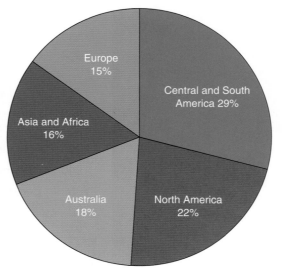

Figure 24.5
Sales of Coca-Cola in different areas of the world

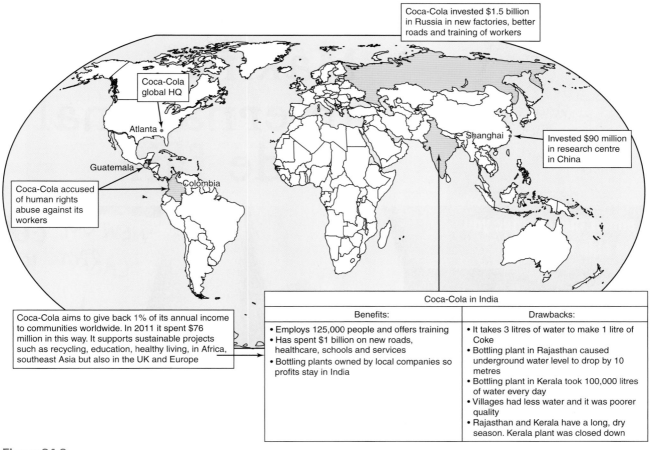

Coca-Cola invested $1.5 billion in Russia in new factories, better roads and training of workers

Coca-Cola global HQ

Atlanta

Guatemala

Coca-Cola accused of human rights abuse against its workers

Colombia

Shanghai

Invested $90 million in research centre in China

Coca-Cola aims to give back 1% of its annual income to communities worldwide. In 2011 it spent $76 million in this way. It supports sustainable projects such as recycling, education, healthy living, in Africa, southeast Asia but also in the UK and Europe

Coca-Cola in India	
Benefits:	Drawbacks:
• Employs 125,000 people and offers training • Has spent $1 billion on new roads, healthcare, schools and services • Bottling plants owned by local companies so profits stay in India	• It takes 3 litres of water to make 1 litre of Coke • Bottling plant in Rajasthan caused underground water level to drop by 10 metres • Bottling plant in Kerala took 100,000 litres of water every day • Villages had less water and it was poorer quality • Rajasthan and Kerala have a long, dry season. Kerala plant was closed down

Figure 24.6
Coca-Cola global developments

Activities

1. Write a report on Coca-Cola as a transnational corporation. Describe:
   (a) how globalised the company is
   (b) the effects it has and has had on other countries and communities.
2. You should include diagrams and maps in your report.

Now complete the 'I can do' boxes for this chapter.

Chapter 25

Making international trade fair

This chapter looks at ways of making international trade fairer.

By the end of this chapter, you should be able to:

✓ describe the advantages and disadvantages of trade alliances
✓ give a definition of *sustainable trade practices*
✓ describe one or more examples of sustainable trade practice.

Figure 25.1

Trade alliances

In Chapter 20 we found out that tariffs and quotas make it very difficult for developing countries to sell their products in developed countries. **Trade alliances are groups of countries between which there is free trade** – there are no tariffs or quotas. They are sometimes called trading blocs.

Trade alliances are most helpful to companies in those countries because they can now sell their goods more easily in the other countries in the alliance. However, **trade alliances have trade barriers for any goods coming from outside** their group, which makes it harder for companies in other countries to sell there.

For example, the European Union (EU) helps companies in the UK, France and Germany to sell goods in Europe, but it makes it much harder for companies from the developing world to do so.

Examples of trade alliances

- The European Union (EU)
- The Economic Community of West African States (ECOWAS)
- North America Free Trade Agreement (NAFTA)

Case study of a trade alliance: The North America Free Trade Agreement (NAFTA)

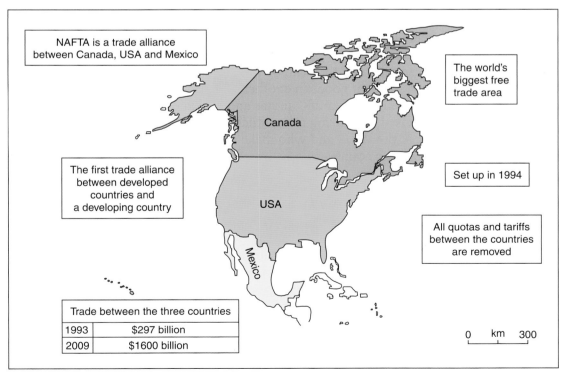

Figure 25.2
The North America Free Trade Agreement

Table 25.1 describes the advantages and disadvantages of NAFTA.

Advantages of NAFTA	Disadvantages of NAFTA
Because there are no trade barriers, trade has increased between the three countries fivefold	More US farm exports to Mexico put Mexican farmers out of business (1.3 million farm jobs lost) because American produce is cheaper (due to subsidies)
This has increased the wealth of all the countries in NAFTA	To compete with American produce, Mexican farmers use more chemicals on the land and cut down trees to make more farmland; this has harmed the environment
US farms especially export more produce	Some companies which moved to Mexico exploit their workers, with very long hours and no employment rights
Mexico and Canada are able to export more oil to the USA	Jobs at US companies increased in Mexico but this caused unemployment in USA and Canada
The USA is now less dependent on the Middle East for oil	
No trade barriers generally mean lower prices for the public	
Many US companies moved to Mexico where wages are lower	

Table 25.1
Advantages and disadvantages of NAFTA

World Trade Organization (WTO)

The World Trade Organization was set up to make international trade fair for every country. All member countries must abide by its rules. The WTO investigates if countries break these rules and they can be punished by trade sanctions (i.e. not being allowed to sell their goods in another country). This has been successful in reducing the number of tariffs, quotas and subsidies in the world. However, the WTO is criticised for making rules which favour rich countries. And of the people who work at the WTO and who make the rules, more are from developed countries than developing countries.

Sustainable practices

Increased trade threatens our environment in many ways – we are over-using our forests, fishing grounds and minerals (see Chapter 22). We are affecting the atmosphere, climate, soil quality and wildlife habitats and because so many goods are needed so fast and so cheaply, in some cases they are being produced in degrading conditions (e.g. people working very long hours, for very low wages, in unsafe conditions). These are all unsustainable practices – the world cannot keep doing this. If it does resources will run out and people's quality of life will suffer. However, **sustainable practices do not threaten the environment or people's quality of life**. Sustainable trade is when goods produced in a sustainable way are traded. Some examples of sustainable trade are shown in the boxes below.

Example 1: Fairtrade

Farmers and producers often receive low prices for their products and these prices fluctuate a lot.

Goods which have the FAIRTRADE Mark have been produced using sustainable practices. Producers such as banana farmers receive a fairer, more stable price and the Fairtrade Premium, which they choose how to invest in their businesses and communities for a sustainable future. We can check that food and even cotton for clothing we buy have been produced in a sustainable way by looking for the FAIRTRADE Mark.

Example 2: Rainforest Alliance Certified™

Large-scale agriculture and forestry are responsible for soil erosion, water pollution, habitat loss and deforestation.

The Rainforest Alliance Certified™ seal is given to all forestry and farm products which are produced in a way that protects the environment. Before we buy paper, timber, chocolate or even a cup of coffee, we can check that it has been produced in a sustainable way by looking for the Rainforest Alliance Certified™ seal.

Example 3: Buying local

Due to globalisation the goods we buy have often been produced thousands of kilometres away and transporting them has a huge impact on the environment.

Goods in shops now have labels showing where they were made or produced. There are more local farmers' markets, farm shops and pick-your-own farms, making it easier for the goods you buy to have a lower carbon footprint.

Other examples of certified products

There are many other schemes that give certificates to companies or products produced in a sustainable way, for example The Soil Association give certificates to all products produced organically, and not just food but health and beauty products too. The Marine Stewardship Council offers certificates for sustainable fishing.

National 4

1. What is a trade alliance? Give one example.
2. Why is it easier for countries to trade with each other in an alliance?
3. Why is it difficult for other countries to trade with those in an alliance?
4. NAFTA is an alliance of the USA, Canada and Mexico. Give one advantage and one disadvantage that NAFTA has brought to Mexico.
5. (a) In what ways are we threatening our environment by increased world trade?
   (b) Why are these practices called 'unsustainable'?
6. Choose one example of sustainable trade. Describe fully how it is better for the environment or people's quality of life.

National 5

1. In what way do trade alliances make trade both fair and unfair?
2. NAFTA is an alliance of the USA, Canada and Mexico. Describe the benefits and disadvantages that NAFTA has brought to

   (a) Mexico
   (b) USA.
3. (a) Describe some unsustainable practices which threaten our environment.
   (b) In what way do unsustainable practices affect people's quality of life?
4. Why is it more sustainable to buy food and goods produced locally?
5. Choose one example of sustainable trade and explain fully how it is sustainable.

25

Activities

Activity A

Look at Guyana's exports and the countries to which it exports.

   (a) Would it help Guyana to form a trade alliance with Canada?
   (b) Do you think Canada wants a trade alliance with Guyana? (Look at its exports.)
   (c) How likely is it that there will be a trade alliance?

Location of Guyana in South America

Activity B

List all the sustainable practices which might make trade fairer for Guyana.

Guyana

Guyana is a small developing country in South America. It finds it difficult to export goods because of trade barriers and has to rely on exporting more and more primary products. This is threatening its environment and its people's quality of life. For example:

- Guyana's resources are being used up.
- The prices Guyana receives are too low for people to have an acceptable standard of living.
- Its farmland is becoming of poorer quality.

Guyana's export partners

- Canada = 28%
- USA = 17%
- UK = 11%
- Ukraine = 6%
- Netherlands = 5%

Guyana's exports

- Sugar
- Gold
- Rice
- Shrimps
- Timber
- Bauxite

Canada's export partners

- USA = 74%
- UK = 4%
- China = 3%
- Japan = 2%
- Mexico = 1%

Now complete the 'I can do' boxes for this chapter.

Chapter 26

This chapter looks at South Korea as a newly-industrialised country.

South Korea – a rapidly changing economy

By the end of this chapter, you should be able to:

✓ describe how employment in South Korea has changed since 1950
✓ give reasons why South Korea has been able to industrialise
✓ give examples of good and bad effects of industrialisation on South Korea.

South Korea

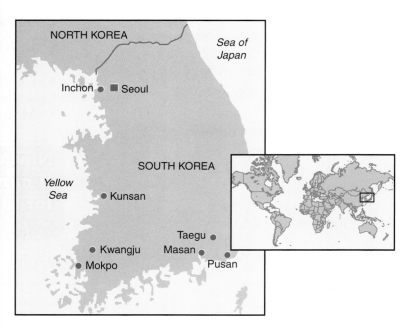

Figure 26.1
Location of South Korea

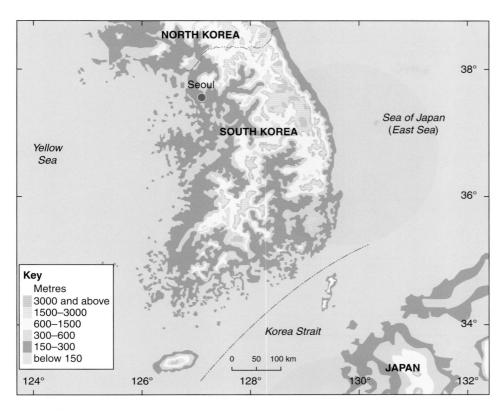

Figure 26.2
Map of South Korea

The country of South Korea was formed in 1953 when it separated from North Korea at the end of the Korean War. It has since transformed itself from a very poor, backward agricultural country to a rich, high-tech industrial giant with one of the biggest economies in the world. This remarkable change is shown in the tables below, in 20-year intervals.

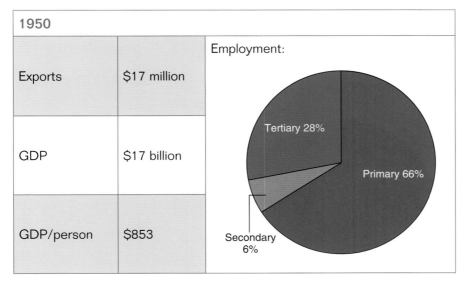

1950		Employment:
Exports	$17 million	
GDP	$17 billion	
GDP/person	$853	

Tertiary 28%
Primary 66%
Secondary 6%

Table 26.1

1970

Exports	$1.1 billion
GDP	$70 billion
GDP/person	$2167

Employment:
- Primary 50%
- Tertiary 36%
- Secondary 14%

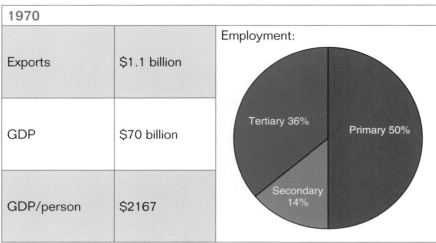

Table 26.2

1990

Exports	$74 billion
GDP	$335 billion
GDP/person	$8704

Employment:
- Primary 18%
- Secondary 28%
- Tertiary 54%

Table 26.3

2010

Exports	$466 billion
GDP	$1519 billion
GDP/person	$29,717

Employment:
- Primary 7%
- Secondary 24%
- Tertiary 69%

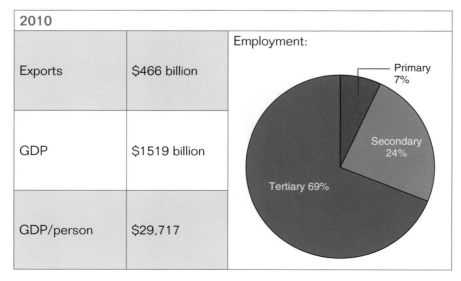

Table 26.4

Industrialisation of South Korea

South Korea began to industrialise in the 1950s, starting its own companies and encouraging companies from abroad to set up there. These companies were attracted to South Korea because of:

- low wages
- low taxes
- an educated workforce.

South Korea protected these new companies with trade barriers. South Korea started making specific products (steel, ships, cars and electronics) which it knew it could sell abroad more cheaply than other countries. It had an especially big market in nearby China. As a result of this industrialisation, the number of people in primary industry decreased and the number in secondary industry increased.

Despite low wages, people were better-off, had more disposable income and could spend money in shops and on entertainment. This meant that many more people found jobs in tertiary industries. The government also had more money to spend on education, research and telecommunications. By 2000, South Korea had one of the best broadband internet systems in the world. It had a very skilled, inventive workforce and was able to start up successful high-tech industries.

Did you know...?
In 1960 South Korea had 100,000 students in higher education. By 1987 it was 1.3 million.

Today, South Korea is the thirteenth biggest economy in the world and is the home of well-known transnational companies. These include car companies such as Hyundai, Daewoo and Kia and electrical companies such as Samsung and LG. It is now the sixth largest producer of motor vehicles in the world and the fourth largest producer of electrical goods. Hyundai is also the biggest shipbuilder in the world. Yet 60 years ago this small country was one of the very poorest in the world.

Figure 26.3
Samsung factory in South Korea

Effects of rapid industrialisation

As the figures in Tables 26.1–26.4 show, in 1950 South Korea exported very little and its exports were mostly primary goods. In 60 years the value of its exports has multiplied by 25,000! Now its exports are mostly secondary goods. South Korea buys and sells from nearly every country in the world – it is truly globalised.

South Korea has become 100 times richer and the average person is now 35 times better off. We can only dream of making this progress in the UK or USA or any European country. People in South Korea now have a much higher standard of education and health care and have the money to afford many luxuries.

But there have been disadvantages. Working conditions have been very poor, especially when the country first began to industrialise. Even now the average working week is 52 hours. Not everyone has benefited from this boom and there is now a much bigger gap between the richest and the poorest in the country. For example, women only get paid 75% of men's wages for doing the same job, although women's rights are generally much improved. There have been a huge number of new buildings, roads and airports constructed which has meant a great increase in air, noise and water pollution. Carbon emissions have tripled in the last 30 years. The proportion of people living in cities has increased from 33% in 1950 to 83% in 2012 and this has led to serious housing shortages. Pollution in the cities is especially bad and smog is common, caused by smoke from factory chimneys and car exhausts.

National 4

1. Look at the four pie charts in Tables 26.1–26.4. Describe the changes in the proportion of people in South Korea working in secondary industry between 1950 and 2010.
2. Draw a line graph showing the changes in:
   (a) GDP/person in South Korea since 1950
   (b) GDP in South Korea since 1950.
   (c) What are the problems of drawing this graph?
3. (a) Name two types of industry in South Korea which began in the 1950s.
   (b) Explain why these industries set up in South Korea.
4. Give two examples which show that South Korea has a very big economy.
5. What do you think has been the biggest benefit to the South Korean people of their country's industrial growth?
6. What has been the biggest drawback?

National 5

1. Look at the four pie charts in Tables 26.1–26.4. Describe the changes in employment in South Korea between 1950 and 2010.
2. Draw a line graph showing the changes in:
   (a) GDP/person in South Korea since 1950
   (b) exports in South Korea since 1950.
   (c) What are the problems of drawing this graph?

National 5 continued...

3. (a) Give examples of the types of industry that have set up in South Korea.
   (b) Give detailed reasons why these industries have set up in South Korea.

4. What do you think has been:
   (a) the biggest benefit to the people of South Korea of its industrial growth? Give reasons for your answer.
   (b) the biggest problem of industrialisation? Give reasons for your answer.

Activities

The events below happened at different times.

   (a) Work out the time order of the events.
   (b) Draw a flow map on a large piece of paper by writing down the events in order. Then put a box around each one and arrows between them. (An example is shown below, with the first event completed.)
   (c) Give your flow map a suitable title.

- South Korean transnational corporations (such as LG) set up in South Wales because it is cheap.
- Steelworks in South Wales close down.
- South Wales is one of the biggest makers of steel in the world.
- UK government offers grants to companies to set up in South Wales.
- Steelworks in South Wales make less and less profit.
- High unemployment in South Wales.
- South Korea begins steelmaking and sells steel abroad.
- Some South Korean TNCs close their factories in South Wales.
- South Wales sells its steel all over the world.

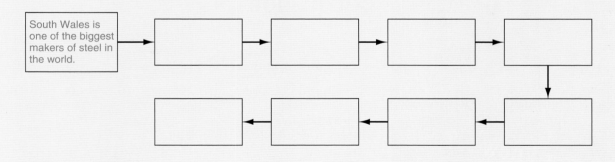

Now complete the 'I can do' boxes for this chapter.

Chapter 27

Health in developing countries

This chapter looks at how healthy people are in developing countries.

By the end of this chapter, you should be able to:

- ✓ describe the main reasons for ill-health in developing countries
- ✓ explain the effects of ill-health on the people and their countries
- ✓ describe some solutions to ill-health in developing countries.

Did you know…?
Six countries in Africa had lower life expectancies in 2010 than in 1970.

Health comparisons between developing and developed countries

	1950	1970	1990	2010
Developing countries	41	52	61	67
Developed countries	66	70	74	78

Table 27.1
Changes in life expectancy (1950–2010)

Table 27.1 shows one of the starkest differences between the worlds of the rich and poor. Health is much better in the rich world and people live far longer. If you are born and live your life in Japan you can expect to live 36 years longer than if you live in the African country of Chad. Of all the inequalities in the world, this is surely one of the worst.

The quality of people's health is used to indicate the level of development of a country; if many people are suffering ill-health, this indicates a low level of development and vice versa. Ill-health makes people very weak and may leave them with a permanent disability, which can result in them quickly becoming trapped in a vicious cycle of disease (see Figure 27.1). In developing countries most people suffer from at least one disease. Not only does this reduce the quality of their lives but it also seriously reduces the economic development of the whole country. This is because:

- people who are unwell cannot work and therefore don't produce wealth
- people who are unwell need other people to look after them – people who otherwise would be working themselves.

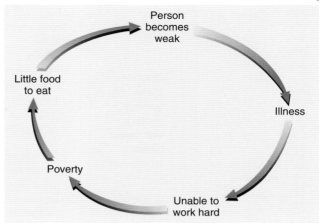

Figure 27.1
Vicious cycle of disease

Causes of ill-health

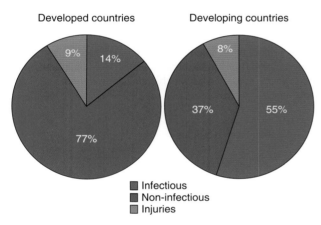

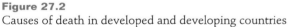

Figure 27.2
Causes of death in developed and developing countries

Did you know...?
Six infectious diseases – pneumonia, tuberculosis, diarrhoea, malaria, measles and HIV/AIDS – cause 90% of all deaths in developing countries.

Diseases can be divided into those that are infectious (where one person infects another) and non-infectious (those which cannot be 'caught' from someone else). In developing countries infectious diseases are far more common and account for most causes of death (see Figure 27.2). Figure 27.3 shows the causes of the most common diseases in developing countries.

Factors in ill-health in developing countries

Drinking **polluted water**.
Lack of proper **sewage disposal** (so the sewage mixes with drinking water).
Climate – warmer climates attract more flies and mosquitoes which spread disease.
Poor health care – to prevent, treat and cure people.
Poor diet – not eating enough food or not eating a balanced diet.
Poverty – people may not have enough money to eat properly or buy medicines and the whole country may not have enough money for proper sewers, water pipes and hospitals.

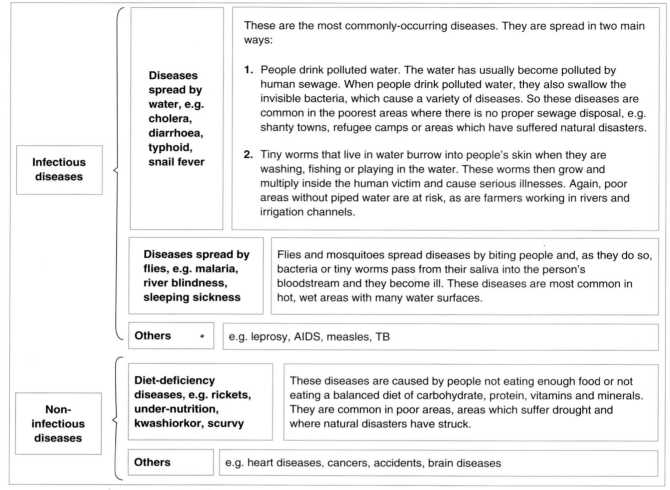

Figure 27.3
Diseases of the developing world

Effects of ill-health

Diseases that are common in the developing world affect people in a variety of ways:

- **Some are killer diseases**, for example AIDS and malaria.
- **Some make people very weak** and lethargic but are not directly responsible for their deaths, for example snail fever and kwashiorkor. If people are very weak, they quickly become trapped in a vicious cycle of disease.
- **Some leave their victims with a permanent disability** or injury, for example river blindness and rickets. Not only does this make it extremely difficult for those affected to find work, but it also requires someone in their community to look after them.
- **Most diseases reduce people's life expectancy**, either directly or indirectly.
- **All diseases reduce the rate of development of the country** and the wealth of the people.

Improving health

Most of the diseases that affect people in the developing world can be prevented or cured. So what are the best strategies for improving health in developing countries?

- **Produce more food** – this reduces diet-deficiency diseases and makes people stronger and better able to fight off diseases.
- **Improve health facilities** – so that everyone, including those living in rural areas, has access to a health centre, trained medical aid and medicines.
- **Provide clean water** – by improving water supplies and sewage disposal.
- **Provide health education** – so people know what causes disease and the simple ways it can be prevented.

National 4

1. Look at Table 27.1.
    (a) What was the difference in life expectancy between developed and developing countries in 2010?
    (b) Is the difference becoming greater or smaller?
2. Which are more common in developing countries – infectious or non-infectious diseases?
3. (a) Name two infectious diseases.
    (b) Name two non-infectious diseases.
4. Describe the vicious cycle of disease shown in Figure 27.1.
5. (a) Explain how polluted water spreads disease.
    (b) Explain how flies spread disease.
6. What causes diet-deficiency diseases?
7. Explain why poverty is an important factor in health.
8. Explain two ways of improving health in developing countries.

National 5

1. Describe what is shown in Table 27.1.
2. Look at Figure 27.2.
    (a) Describe the differences in cause of death between developed and developing countries.
    (b) Explain fully how polluted water spreads disease.
    (c) Explain fully how flies spread disease.
3. What are the main causes of non-infectious diseases in developing countries?
4. 'Poverty causes ill-health and ill-health causes poverty.' Explain what this statement means.
5. Which diseases would be reduced by providing clean water to a community?

Activities

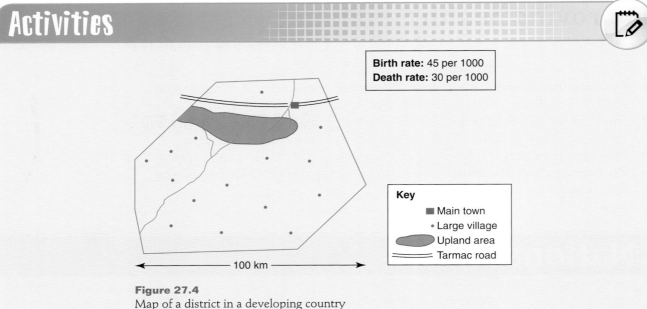

Birth rate: 45 per 1000
Death rate: 30 per 1000

Key
- ■ Main town
- • Large village
- ⬭ Upland area
- ═ Tarmac road

← 100 km →

Figure 27.4
Map of a district in a developing country

Activity A

The government of a developing country has made health a priority. It has set aside an extra amount of money for its very poorest farming district which has the lowest life expectancy in the country (see the above figure). The government knows there are three big health problems but it is unsure how to tackle them. The problems and possible solutions are shown below.

1. Diseases spread by polluted water. *Either:* install sewers so that sewage does not mix with the water supply, *or:* provide free treatment to people suffering from these diseases.
2. Poor diet. *Either:* install small reservoirs (tanks) which allow people in each village to store water, *or:* give a free meal daily to everyone.
3. Lack of health facilities. *Either:* build a small health centre in every large village (with a nurse who can give health advice and provide cheap medicines), *or:* build one large hospital in the biggest town (with doctors, better equipment, surgery and maternity facilities).

For each of the three problems, decide which of the two solutions you think is best. Justify your answer.

Activity B

Unfortunately the government has less money than it expected and can only afford to tackle one of the above problems. Which one should it tackle, and why? (Think what health problems each would solve and the knock-on effects of solving these.)

Now complete the 'I can do' boxes for this chapter.

Chapter 28

Health in developed countries

This chapter looks at how healthy people are in developed countries.

By the end of this chapter, you should be able to:

✓ explain why developed countries have less disease than developing countries
✓ describe the areas within developed countries which have the worst health
✓ describe the main factors affecting health in developed countries.

Health in developed countries

As mentioned in Chapter 27, people in developed countries can expect to live on average 11 years longer than people in developing countries. There are three main reasons for this:

1. Clean environmental conditions
2. High-quality health facilities
3. Improved health education

Table 28.1 compares various indicators of health between developed countries and developing countries.

Clean environmental conditions

People in developed countries live in a cleaner, healthier environment. Our water is purified before it reaches our taps. Sewage is taken away by pipes and treated before being emptied into rivers and seas. Rubbish is collected regularly and disposed of properly. Under these sanitary conditions, infectious diseases are much less likely to spread.

	Gross national income per person ($)	% people with improved sanitation	Doctors per 10,000 people	Calories per person per day
Norway	88,890	100	42	3460
Australia	49,130	100	30	3190
Canada	45,560	100	20	3530
Japan	44,900	100	21	2810
Ireland	39,930	100	32	3530
Pakistan	1120	48	8	2250
Kenya	820	32	2	2060
Haiti	700	17	3	1850
Ethiopia	370	21	0.2	1950
DR Congo	190	18	1	1590

Table 28.1
Effects of a country's income on factors affecting health

High-quality health facilities

Wealthier countries can afford to spend a lot of money on medical care. A full range of health care is available, including surgical operations, organ transplants and antenatal care, provided by a range of health personnel including doctors, nurses, midwives and physiotherapists. Children are routinely inoculated against diphtheria, tetanus and polio when they are a few months old; against measles, mumps and rubella when one year old, and against tuberculosis when about 13 years old. To ensure that people are not affected by the vicious cycle of disease, sickness benefit is paid to people who are unable to work through illness.

Improved health education

People in the developed world are much more aware of the causes of disease and how they can be prevented. We know the importance of a healthy diet, regular exercise and safe sex. It is much easier to inform people of health matters in richer countries. Countless radio programmes and TV chat shows discuss topical health issues. Health services regularly run campaigns and they get their message across through advertisements in newspapers, on television, in schools and on roadside billboards.

Ill-health in developed countries

Wealthier countries have been very successful in reducing infectious diseases, such as whooping cough, diphtheria, scarlet fever and tuberculosis. We have learned how to prevent and cure these diseases and they now account for relatively few deaths. Instead, it is the non-infectious diseases such as heart disease, diabetes and cancer that are more serious in developed countries. These are more difficult to prevent and cure, mostly because we do not fully understand their causes.

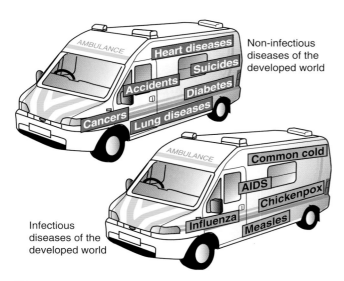

Figure 28.1
Examples of infectious and non-infectious diseases of the developed world

Five factors are particularly important in explaining the distribution of non-infectious diseases in developed countries:

1. Pollution
2. Social habits
3. Poor diet
4. Stress
5. Access to health facilities.

Pollution

Air pollution from vehicles, factories and power stations is more serious in developed countries. Breathing in polluted air can cause lung disease and some forms of cancer. These diseases are more common in:

- urban areas rather than rural areas
- areas with a high concentration of heavy industry
- countries where environmental laws are less strict.

Social habits

Social habits such as smoking and drug and alcohol abuse increase our chances of dying from a variety of diseases, such as lung cancer and liver disease. Almost 3000 people in Scotland died from alcohol-related causes in 2012. Research also shows that smokers die ten years earlier on average than non-smokers. These habits are expensive and so are more common in richer countries but, within developed countries, the highest rates of smoking and alcohol-related illnesses are found in more deprived areas.

Poor diet

Unlike in the developing world where people suffer from a lack of food, in developed countries people suffer ill-health from eating too much, especially fatty foods. This puts us at risk of heart disease and cancer. Research has shown that 30% of children and 65% of adults in Scotland are overweight and that Scottish children are more inactive, unfit and overweight than ever before. As a result, their life expectancy may be less than that of their parents. Cheaper foods often contain more fat so, within developed countries, ill-health due to a poor diet is also more common in poorer areas.

Stress

The faster, more hectic pace of life in developed countries affects our health. Stress is linked to heart disease, brain diseases, even accidents and suicides. Generally, stress levels are greater for those living in cities than in the countryside.

Figure 28.2
Poor diet is often a cause of disease in developed countries

Access to health facilities

Within the developed world there are significant differences in health facilities available to people in different countries and different parts of a country. For example, the USA spends four times as much per person on health than Spain, and 100 times more than Poland. But in many developed countries, including the USA, health care has to be paid for and, as a result, life expectancy is lower in poorer areas. A report by the NHS in 2011 found that life expectancy for men in Lenzie, just outside Glasgow, was 82 years while in the Calton area of Glasgow it was only 54 years.

National 4

1. Name four non-infectious diseases more common in developed than developing countries.
2. Why are infectious diseases less likely to spread in developed countries?
3. Describe how developed countries protect their children from catching diseases.
4. In what ways are people in developed countries educated about health?
5. Five main factors that cause ill-health in developed countries are pollution, stress, social habits, poor diet and access to health facilities. Which of these affect people in:
    (a) urban areas more than rural areas?
    (b) poorer areas more than richer areas?

National 5

1. (a) Name three infectious and three non-infectious diseases common in developed countries.
   (b) Which account for more deaths – infectious or non-infectious diseases?
2. Why are infectious diseases less likely to spread in developed countries?
3. Explain how we are able to prevent diseases in the developed world more easily than in the developing world.
4. Life expectancy in cities in developed countries is lower than in the countryside. Explain why.
5. Life expectancy is lower in poorer areas of developed countries. Explain why.

Activities

Activity A

Pollution, social habits, poor diet and stress are four main causes of ill-health and death in rich countries.

(a) Which do you think is the biggest cause? Why?

(b) Which do you think is the easiest to reduce? How would you do it?

Activity B

A

B

Health messages encouraging the use of seat belts (the Clunk-Click seat belt campaign) and to use Sun protection to prevent overexposure to the Sun (the Slip! Slop! Slap! Sun protection campaign)

Give examples of health messages you can remember.

(a) How were they advertised (e.g. on TV, on posters)?

(b) How effective do you think they were?

Now complete the 'I can do' boxes for this chapter.

Chapter  29

Malaria – its cause and transmission

By the end of this chapter, you should be able to:

- ✓ describe the distribution of malaria around the world
- ✓ explain the causes of malaria
- ✓ describe how malaria is transmitted.

Did you know...? Every 60 seconds, a child dies from malaria.

Malaria

Infectious diseases are more common in poorer countries and non-infectious diseases are more common in rich countries. To show the effects of these diseases, the most common ones affecting both developing and developed countries will now be studied in more detail. The first disease we will look at affects 400 million people and kills 1.2 million every year, half of them children. This disease is called **malaria**.

Cause and method of transmission of malaria

Malaria is caused by a tiny parasite that finds its way into a person's bloodstream. After a few days, the infected person suffers headaches and stomach pains, followed by fevers of high temperature and shivering fits. The fevers can occur many times, frequently resulting in the death of the victim. Malaria is a particularly big killer of children, who have not had time to build up any immunity to the disease. Even if malaria does not kill the victim, it can cause kidney failure and leaves the patient weak, anaemic and prone to other diseases. The infected person's life expectancy is reduced considerably.

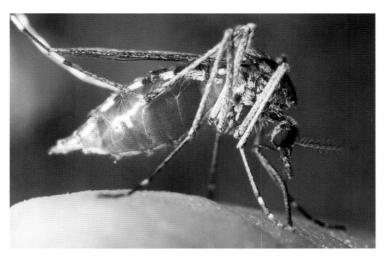

Figure 29.1
Mosquito sucking up blood

In areas where malaria occurs, many of the people will have the disease. As a result, the amount of wealth (from farms and factories) that the area produces is seriously reduced while, at the same time, a lot of time and money has to be spent on caring for victims of the disease. In the Philippines, for example, malaria was rife in the 1940s and as a result 35% of people were unable to work. In regions where malaria is particularly bad, for example northern Sri Lanka, people have been forced to move away, leaving behind fertile farmland.

The parasite that causes malaria enters a person's bloodstream when he or she is bitten by a mosquito. Not all mosquitoes carry the disease. **Only the female *Anopheles* mosquito spreads malaria.** It bites an infected person and sucks blood containing the parasite into its stomach, where the parasites multiply. The mosquito then bites someone else and the parasite enters the new victim via the saliva of the mosquito.

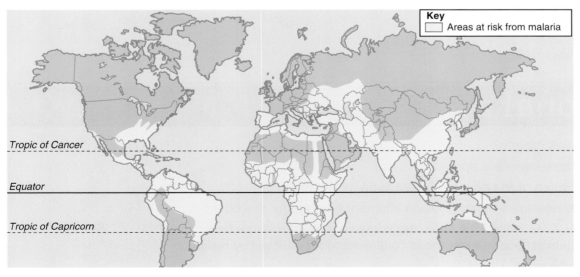

Figure 29.2
Distribution of malaria (1940s)

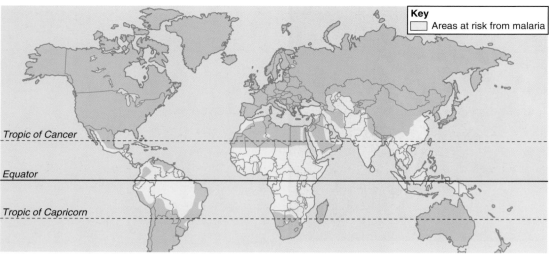

Figure 29.3
Distribution of malaria (1990s)

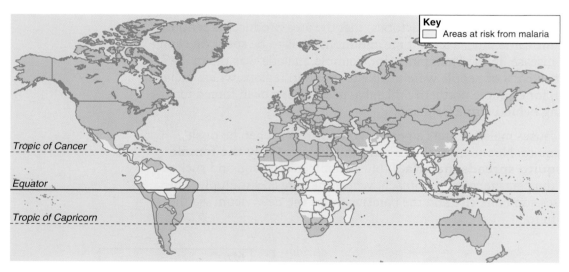

Figure 29.4
Distribution of malaria (2010s)

National 4

1. What causes malaria?
2. Describe how malaria is spread.
3. What are some of the effects of malaria on its victims?
4. Give three examples of how malaria affects the economy of a country.
5. Figures 29.2, 29.3 and 29.4 show the spread of malaria in three decades – 1940s, 1990s and 2010s. In which decade were these continents most affected by malaria?
   (a) North America
   (b) Africa
   (c) Asia
6. Overall, how has the global distribution of malaria changed since the 1940s?

National 5

1. Describe in detail the causes of malaria and how it is spread.
2. Give examples of some of the effects of malaria on its victims.
3. Explain how malaria slows down the economic development of a region.
4. Look at Figures 29.2 and 29.3. In which continents had malaria increased between the 1940s and 1990s, and in which continents had it decreased?
5. Look at Figures 29.3 and 29.4. In which continents had malaria increased between the 1990s and 2010s, and in which continents had it decreased?
6. Look at Figure 29.4 and describe the global distribution of malaria in the 2010s.

Activities

Country	1940s	1990s	2010s
Australia	SOME	NONE	NONE
Cuba			
Italy			
Kazakhstan			
Madagascar			
Paraguay			
Somali Republic			
Spain			
South Korea			
Turkey			
Venezuela			
Yemen			

Table 29.1
Countries with malaria

(a) Copy Table 29.1 into your notebook.
(b) Using an atlas and Figures 29.2, 29.3 and 29.4, complete the table. For each time period, you must write down whether ALL, SOME or NONE of the country had malaria. The first one has been done for you.

Now complete the 'I can do' boxes for this chapter.

Chapter 30

This chapter looks at the factors affecting malaria.

Malaria – factors in its distribution

By the end of this chapter, you should be able to:

✓ describe the physical environment where malaria is likely to be found
✓ explain the human factors that contribute to the spread of malaria
✓ give examples of methods used to control malaria.

Factors in the distribution of malaria

Malaria has occurred in most areas of the world at some time. As Figure 29.2 shows, in the 1940s even people in the USA and Europe were affected. The disease then retreated (Figure 29.3) and the most recent map of its distribution (Figure 29.4) shows little change from the 1990s. Today, malaria exists in 100 countries and over 60% of the world's population live in malaria-affected areas. The reasons behind its changing distribution are a combination of physical and human factors.

Physical factors

Malaria occurs in areas where the *Anopheles* mosquito lives. These mosquitoes live in both warm and hot areas, where **temperatures are above 16°C**. They need **still water surfaces** as breeding areas, but these areas do not need to be large. As a result, all **warm, rainy areas** with still or slow-moving water are suitable environments for the *Anopheles* mosquito.

Human factors

People's activities also affect the distribution of malaria. **Where people have built dams and made irrigation channels**, they have created suitable breeding areas for the mosquito and so the incidence of malaria increases. **People migrate much more now** and this makes it easier for the disease to spread. In some areas of the world, people have successfully used insecticides to kill off the mosquito. This explains why the areas affected by malaria decreased between the 1940s and the 1990s.

Controlling malaria

Malaria can be both prevented and cured by anti-malarial drugs such as Chloroquine, Larium and Malarone, which kill the parasite that causes the disease. However, the malarial parasites can develop resistance to anti-malarial drugs, so the drugs have to be constantly re-developed in order to be effective. Instead of curing people, it is much better and cheaper to prevent people from catching malaria by reducing their exposure to mosquitoes. **To prevent malaria, the mosquitoes must be destroyed.** Many methods of prevention have been tried (see Table 30.1) but they all have drawbacks, which is why so little progress has been made in recent years.

Did you know....?
Mosquitoes bite mostly at night, which is why mosquito bed nets are so important.

Did you know....?
Malaria is preventable, treatable and curable, yet it is the largest killer of children in the world.

Method of prevention	Drawbacks
Use insecticides (e.g. DDT and Malathion)	Some chemicals pollute the environment, killing other life forms Mosquitoes are becoming resistant to some insecticides Some insecticides are expensive
Use insecticide-treated bed nets to prevent being bitten at night	Developing countries cannot afford to buy bed nets
Drain breeding grounds	Impossible to drain all breeding grounds, as only small areas of water are needed for larvae to exist, e.g. potholes in roads
Kill the mosquito larvae (e.g. using egg whites, coconuts, mustard seeds and fish in the water)	This is wasteful of food; it is also wasteful of water
Educate people, so they know how the disease is spread and can reduce their contact with mosquitoes	People still have to live near water, which is where mosquitoes live

Table 30.1
Methods of preventing malaria

National 4

1. Describe the physical environment in which *Anopheles* mosquitoes live.
2. In what ways have people helped to spread malaria?
3. Describe the ways in which malaria can be prevented.
4. Choose one of the methods of prevention shown in Table 30.1 and explain whether you think it is effective or not.
5. Describe how malaria can be cured.

National 5

1. Describe, in detail, the physical environment in which malaria is found.
2. Explain how people have helped to spread malaria.
3. Describe in detail the ways in which malaria can be prevented.
4. Look at Table 30.1 Which are the two best methods of preventing malaria? Explain your decision.
5. Explain in detail why little progress has been made in recent years in reducing the spread of malaria.

Activities

Activity A

Study the image below carefully.

(a) Do you think the people in the photo suffer from malaria?
(b) What other information would help you to answer this question?

Activities continued...

Activity B

Imagine you are a voluntary worker who has gone to a village to warn the people living there of the dangers of malaria and to explain how they can reduce the risk of contracting the disease. Make an information poster informing the villagers of what they can do to reduce their chances of contracting the disease.

Activity C

'There is much more research into male baldness than there is into diseases such as malaria.'

(Bill Gates)

Look at the statement above by Bill Gates, Co-founder of Microsoft. Write down what you think about this statement.

Now complete the 'I can do' boxes for this chapter.

Chapter  31

This chapter looks at a common disease in rich countries – heart disease.

By the end of this chapter, you should be able to:

✓ describe the main causes of heart disease
✓ draw and interpret graphical data.

Heart disease – its causes

Heart disease

In developed countries, heart disease is the second biggest cause of death after cancer. It kills nearly half of all men and women. One in four men will have a heart attack before retirement age and most teenagers show signs of narrowing of the blood vessels, which is the start of heart disease. Unlike major diseases in developing countries, heart disease is non-infectious – one person cannot infect another. Instead, the causes of heart disease are more complicated.

Causes of heart disease

Heart (cardiovascular) disease can cause stroke, angina and heart attacks. Some types of heart disease affect the arteries (which carry blood from the heart to the rest of the body); others affect the heart itself. Many factors contribute to heart disease.

Poor diet

Too many fatty foods increase cholesterol, which is a type of fat found in the blood. This narrows the arteries, increasing the chance of heart disease. Fatty and sugary foods also lead to a person becoming obese or overweight, which puts an extra strain on their heart.

Lack of exercise

Lack of exercise raises blood pressure and cholesterol levels and can also cause a person to become overweight.

Smoking

Nicotine increases the heart rate and blood pressure, so more oxygen is needed for the heart to work properly. However, smokers receive less oxygen while smoking, putting their heart under extra strain. A packet of cigarettes a day doubles a person's chances of having a heart attack and makes them five times as likely to have a stroke.

Stress

Stress increases a person's blood pressure and this puts extra pressure on their heart. People under stress often indulge in 'comfort eating', for example chocolate bars or greasy chips, which can also cause heart disease.

Inheritance

People can inherit a risk of high blood pressure and high cholesterol levels from their parents.

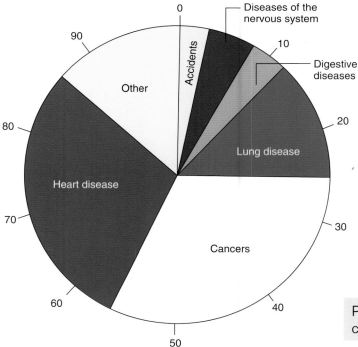

Pie-charts should be used to compare parts of a whole.

Figure 31.1
Causes of death in England and Wales (2010)

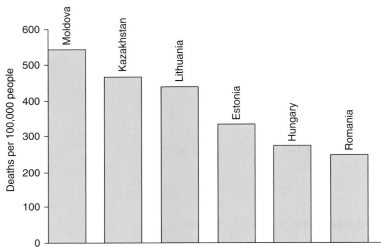

Bar charts should be used to compare separate amounts.

Figure 31.2
Countries with the highest death rate from heart disease (2008)

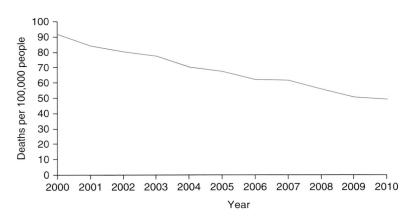

Line graphs should be used to show changes over distance or time.

Figure 31.3
Changes in death rate from heart disease in Scotland (2000–2010)

Type of cost	Amount (£ billion)
Prevention	0.3
Primary care (e.g. work done by GPs)	1.1
Outpatient care at hospitals	1.0
Inpatient care (hospital admissions)	4.3
Medicines	2.0
Rehabilitation (recovery needs)	0.5
Social services (care at home)	1.6

Table 31.1
Cost of health care for heart disease in the UK (2009)

Type of cost	Amount (£ billion)
Health care	8.4
Productivity loss (loss of production because people cannot work or die young)	3
Other care costs (e.g. family members looking after those with heart disease)	2.7

Table 31.2
Total cost of heart disease in the UK (2004)

City	Death rates from heart disease per 100,000 people			
	2004	2006	2008	2010
Aberdeen	123	106	85	68
Dundee	134	117	104	100
Edinburgh	116	113	128	131
Falkirk	119	107	98	84
Glasgow	148	133	119	107
Stirling	128	111	113	97

Table 31.3
Death rates from heart disease in Scottish cities (2004–2010)

National 4

Cause of heart disease	Explanation of the cause
Poor diet	
Lack of exercise	
Smoking	
Stress	
Inheritance	

1. Copy the table above into your notebook and complete it using the information in this chapter.
2. Look at Figure 31.1. What percentage of people in the UK die from:
    (a) heart disease
    (b) cancer?
3. Look at Figure 31.2. What is the death rate from heart disease in:
    (a) Moldova
    (b) Romania?
4. Look at Figure 31.3 and describe the changes in deaths from heart disease in Scotland since the year 2000.

National 4 continued...

5. Look at Table 31.3.
   (a) Between 2004 and 2010, in which city did death rates from heart disease:
      (i) fall the most
      (ii) increase the most?
   (b) Draw the most appropriate graphs (line, bar or pie) to show these changes.
6. What would be the most appropriate graph (line, bar or pie) to show the information in Table 31.1? Give a reason for your answer.
7. Table 31.2 shows the costs of heart disease to the UK economy. Choose one of the items in this table and give details of the costs involved.

National 5

1. Describe the effects of smoking on your heart.
2. Explain how eating fatty foods increases your chances of heart disease.
3. Explain how obesity can lead to heart disease.
4. Heart disease can be caused by high blood pressure. What can cause people to have high blood pressure?
5. Describe what is shown by Figure 31.1.
6. Look at Figure 31.2. Compare the death rates from heart disease in Moldova, Hungary and Romania.
7. Look at Figure 31.3. Describe in detail the changes in deaths from heart disease in Scotland since the year 2000.
8. Explain what is shown by Table 31.2. You may wish to use the information in Table 31.1 in your answer.
9. Look at Table 31.3.
   (a) Which three cities reduced the death rate from heart disease the most between 2004 and 2010?
   (b) Draw an appropriate graph(s) to show these changes.
10. What would be the most appropriate graphs (line, bar or pie) to show the information in Tables 31.1 and 31.2? Give reasons for your answer.

Activities

> Scotland has been branded the heart attack capital of Western Europe, caused largely by

> Cases of heart disease finally starting to fall in Scotland. Doctors put it down to

> Research claims heart disease was much lower in Scotland 100 years ago. Scientists say this was due to

> New report says it should be easy to reduce heart disease in Scotland. We just need to

The above newspaper headlines have been torn off. Write them out and complete what you think they were going to say.

Now complete the 'I can do' boxes for this chapter.

Chapter 32

Heart disease – methods of control

This chapter looks at the distribution of heart disease and ways of controlling it.

By the end of this chapter, you should be able to:

- ✓ describe the factors in the distribution of heart disease
- ✓ give examples of methods of controlling heart disease
- ✓ explain the role played by the National Health Service in preventing and treating heart disease.

Factors in the distribution of heart disease

The countries most affected by heart disease are all developed countries, but there are big differences between them. Scotland has one of the worst heart disease rates in the world, but even here, some areas (e.g. Glasgow) are much worse than other areas. The main reasons why heart disease varies so much from one area to another and from one developed country to another are described below.

Lifestyle

In developed countries and in cities throughout the world, **the pace of life is faster**. Many more people work in offices and take little exercise.

Diet

Some countries have healthier diets than others. For instance, the traditional Asian diet contains very little meat and dairy produce and Japanese people have a much lower heart disease rate than British people. A Mediterranean diet also contains fewer saturated fats and people there have lower rates of heart disease too.

Affluence

People in richer countries can afford to eat too much and can afford to buy cigarettes and alcohol, so are more likely to develop heart disease. Within richer countries, **the cheapest foods are often fatty foods** and so poorer people are more likely to develop heart disease.

Medical care

Developed countries, for example Australia, have run very good **campaigns to educate people on how to reduce heart disease**. As a result, the incidence of heart disease has reduced dramatically in recent years.

Within any country, **the number of people with heart disease depends on the treatment available** locally, for example, how well the local health authorities try to diagnose and prevent heart disease and the equipment and drugs available there to treat the disease.

Controlling heart disease

The death rate from heart disease in the UK has been dropping in recent years; it dropped by 44% between 2000 and 2010. Some countries, such as Australia, Canada and Sweden, have done even better than this. The death rate has dropped because of better prevention and better treatment.

In the UK, **the National Health Service (NHS) helps to control heart disease** by providing free medical check-ups, giving advanced treatment and by educating people.

Medical check-ups

More people now have regular cholesterol and blood pressure check-ups, which enables them to find out if they are at risk of heart disease and to take action before it is too late.

Advanced treatment

Better medical equipment is being invented and used in the treatment of heart disease, including pacemakers, artificial heart valves and defibrillators. The success rate of heart by-pass surgery is steadily improving and more drugs are being developed, for example aspirin to reduce blood clotting, beta-blockers to reduce heart rate and alpha-blockers to reduce blood pressure.

Education

People are being educated through campaigns in the media and on posters. Money spent on preventing people developing heart disease is money well spent, as the cost of treating them for heart disease is much greater. The four main pieces of advice given by the NHS are shown below.

Eat more	Eat less
skimmed milk	full milk and cream
polyunsaturated margarine	butter
grilled food	fried food
low-calorie soft drinks	milk shakes
chicken, turkey	sausages, pork pies
oats, pasta, cereals	cakes, biscuits, sweets
fruit and vegetables	chips, crisps
brown bread	white bread

Figure 32.1
Preventing heart disease

1. **Eat a better diet.** Advice is given on foods that are healthy and which foods should be reduced or avoided (see Figure 32.1). Food labels now contain much more information. Despite these measures, the number of overweight and obese children in Scotland is increasing, from 27% in 2000 to 30% in 2010.
2. **Take more exercise.** People are encouraged to take more exercise and sports facilities have increased, for example there are now more jogging tracks, cycle lanes, gyms and sports centres. As a result, evidence suggests that the average person now takes more exercise.
3. **Stop smoking.** There have been extensive campaigns to persuade people to stop smoking, and there is more help available than ever before, for example nicotine patches, helplines and hypnotism. The number of smokers is now fewer than 20 years ago.
4. **Reduce stress levels.** People are now more aware that stress is harmful and understand ways to reduce stress, for example relaxing by taking exercise or listening to music. But there is no evidence that stress levels are decreasing. In fact, it is more likely that they are increasing.

National 4

1. Name four factors affecting the distribution of heart disease.
2. How does lifestyle affect the distribution of heart disease?
3. How does affluence affect the distribution of heart disease?
4. (a) What advice is given to people to prevent heart disease?
   (b) Are people in the UK taking this advice?
5. How does the NHS try to control heart disease?

National 5

1. Describe in detail three factors that affect the distribution of heart disease.
2. Explain clearly the ways that heart disease could be prevented.
3. The NHS tries to control heart disease by educating people.
   (a) Describe the advice people are given.
   (b) Describe how successful this advice is.
   (c) Describe two other ways in which the NHS controls heart disease.

Activities

Activity A

Health indicator	Year 2000	Year 2010
Smokers (adults)	28%	21%
Taking enough physical exercise (male)	36%	42%
Taking enough physical exercise (female)	23%	31%
Heavy drinkers (female)	10%	13%
Heavy drinkers (male)	21%	19%
Low-calorie soft drinks (ml per person per week)	516	579
Skimmed milk (litres per week)	1.16	1.16
Sugar (grams per person per week)	130	90
Salt (grams per person per week)	9	11
Butter (grams per person per week)	37	40
Low-fat spread (grams per person per week)	22	11
Chocolate bars (grams per person per week)	113	89
Fruit and vegetables (grams per person per week)	2240	2381

Table 32.1
Comparison of health indicators in 2000 and 2010

You have been asked to give a report to the Minister of Health on how healthily we live in the UK.

(a) Look at Table 32.1. Using bullet points, list the four main points that you would make to the Minister.

(b) Describe your main recommendations. What are the two main problems and how do you think they would best be solved?

Activities continued...

Activity B

Choose one of the methods of controlling heart disease listed below.

(a) Write down what you think would be the best way of persuading teenagers to adopt the method and why.

(b) Draw a poster with an eye-catching slogan to get this message across.

- Eat a better diet
- Stop smoking
- Take more exercise
- Reduce stress levels

Now complete the 'I can do' boxes for this chapter.

Chapter 33

This chapter looks at HIV/AIDS.

HIV/AIDS – its distribution

By the end of this chapter, you should be able to:

✓ explain how HIV/AIDS affects people
✓ describe the global distribution of HIV/AIDS
✓ draw and interpret choropleth maps.

Did you know...?
HIV/AIDS is the world's leading infectious killer.

Did you know...?
In Africa, because of HIV/AIDS, life expectancy is now lower than it was 30 years ago.

HIV/AIDS

AIDS (acquired immunodeficiency syndrome) is caused by a virus called HIV (the human immunodeficiency virus). If you are infected with HIV, your body tries to fight the infection by making antibodies. A person who is HIV-positive has these antibodies inside them, which means they have the HIV virus.

The HIV virus gradually wears down a person's immune system, making it more and more difficult for them to fight disease. This means that viruses and bacteria (which do not cause ill-effects in healthy people) make people who are HIV-positive very sick. They develop a group of health problems (a syndrome) called AIDS and eventually, because they have little resistance to disease, HIV-positive people will die.

Distribution of HIV/AIDS

In 2010 there were 34 million people worldwide living with HIV/AIDS. Every country in the world has cases, but the overall distribution is very uneven (see Table 33.1).

Two-thirds of all cases of HIV/AIDS are in Africa but the distribution there is also very uneven. Figure 33.2 is a choropleth map showing the cases of HIV/AIDS in Africa. The countries are shaded differently according to the total number of cases found there – the darker the shading, the greater the number of cases. The map clearly shows the uneven distribution.

Region (see Figure 33.1)	Number of people with HIV/AIDS (million)	Percentage of all adults with HIV/AIDS (%)
Sub-Saharan Africa	22.9	5.0
North Africa and Middle East	0.47	0.2
South and Southeast Asia	4.0	0.3
East Asia	0.79	0.1
Oceania	0.05	0.3
Latin America	1.5	0.4
Caribbean	0.20	0.9
Eastern Europe and Central Asia	1.5	0.9
North America	1.3	0.6
Western and Central Europe	0.84	0.2
WORLD TOTAL	34	0.8

Table 33.1
People living with HIV/AIDS (2010)

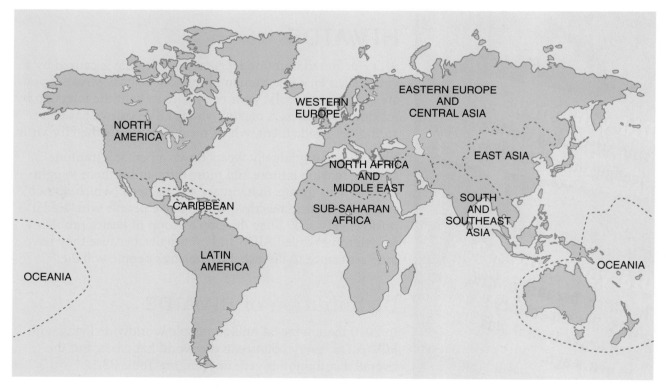

Figure 33.1
Regions of the world referred to in Table 33.1

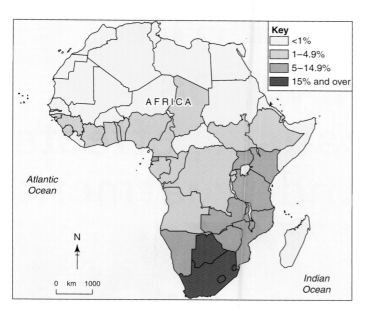

Figure 33.2
Percentage of people affected by HIV/AIDS in African countries, 2012

National 4

1. Describe how HIV affects its victims.
2. Look at Table 33.1.
   (a) Which two regions have the most cases of HIV/AIDS?
   (b) How many people have HIV/AIDS in these two regions?
   (c) How many people in the world have HIV/AIDS?
3. Look at Figure 33.2 and a map of Africa.
   (a) Name three countries where HIV/AIDS is less than 1%.
   (b) Name three countries where HIV/AIDS is greater than 10%.
4. (a) Mark the regions shown in Figure 33.1 on an outline map of the world.
   (b) Draw a choropleth map to show the percentage of adults with HIV/AIDS in these different regions, using the categories given by your teacher.

National 5

1. Describe in detail how HIV/AIDS affects the victim.
2. Using Table 33.1, describe the global distribution of HIV/AIDS.
3. (a) Mark the regions shown in Figure 33.1 on an outline map of the world.
   (b) Draw a choropleth map to show the percentage of adults with HIV/AIDS in these different regions. You will first need to decide how many categories you wish to have and what the categories will be.

Now complete the 'I can do' boxes for this chapter.

Chapter 34

HIV/AIDS – causes, effects and treatment

This chapter looks at factors affecting the distribution of HIV/AIDS.

By the end of this chapter, you should be able to:

✓ describe the factors affecting the distribution of HIV/AIDS
✓ explain the consequences of HIV/AIDS for a country
✓ give examples of how HIV/AIDS can be treated.

Factors in the distribution of AIDS

When the blood or body fluids of a person with HIV/AIDS are passed on to someone else, that person also becomes infected. **The main ways in which people contract HIV/AIDS are:**

- **sharing a needle with an infected person**
- **having unprotected sex with an infected person**
- **babies drinking the breast milk of an infected mother.**

Once a person is HIV-positive, other factors which weaken a person's immune system make him or her more likely to develop AIDS. **The main weakening factors are:**

- **drug abuse**
- **poverty**
- **malnutrition**
- **depression**
- **other infections.**

In poorer countries, especially in Africa, the greater number of cases of AIDS is due to the greater number of people living in poverty and suffering from malnutrition and other infections. In addition, there is less health education so many people are unaware of the causes of AIDS and how the risk of infection can be reduced. **War and the breakdown of law accelerate the spread of AIDS** because of the poverty and malnutrition which follows, and because of the higher incidence of rape. Sadly, many areas of sub-Saharan Africa have suffered conflict in recent years, for example Sudan and Democratic Republic of the Congo.

Did you know....?
More than 6800 new cases of HIV occur each day worldwide.

In developed countries, drug abuse is a much bigger factor in the spread of AIDS than it is in developing countries.

Consequences of HIV/AIDS

In 2010, 69% of the people with HIV/AIDS lived in Africa and in some African countries one in every three adults was infected. The consequences for these countries are extremely serious.

- **The prevention, detection and treatment of AIDS is expensive.** To treat all the AIDS-infected people in Africa would cost three times the amount of money available for all health care. In some African countries, over one-third of all hospital beds are now occupied by HIV/AIDS patients. **To effectively treat AIDS means less money and less health care for people with other diseases.**
- As adults become ill, **responsibility falls on the older children to try and earn money, provide food and care for their family**. This is almost impossible for them and is often at the expense of their own education. So **the next generation of African adults will be less educated, less wealthy and less healthy** than the previous generation.
- Table 34.1 shows the effect of AIDS on life expectancy in a sample of African countries. **People are dying younger** and so have fewer years when they are economically active. In addition, as they fall ill **they are less able to work.** With so many adults affected, **this is seriously reducing production on farms, in factories and in offices** in every African country. South Africa has calculated that its total income will reduce by nearly 20% because of AIDS. Because there are fewer people working, **there are fewer taxpayers,** so **the country is producing less wealth** and also has less tax money to pay for services and to carry out development plans.

Treatment

There is **no cure for AIDS** but there are ARV (antiretroviral) **drugs which slow down the effects of the HIV virus**. There are also **drugs which stop the disease passing from pregnant mothers to their babies. Health education programmes** can be used to try to prevent the disease from spreading. It is also important for people to be tested, as many do not know they are carrying the disease. In addition, efforts can be made to **reduce the effects of factors which hasten the development of AIDS**. Attempts to reduce poverty and improve diet should therefore reduce malnutrition and slow down the progress of the disease.

Country	Life expectancy before AIDS	Life expectancy in 2010
Botswana	74.4	53.3
Lesotho	67.2	46.0
Malawi	69.4	51.5
Namibia	68.8	61.9
Rwanda	54.7	53.9
South Africa	68.5	51.2
Swaziland	74.6	47.3
Zambia	68.6	46.9
Zimbabwe	71.4	46.5

Table 34.1
Changes in life expectancy in nine African countries

34

Case Study: South Africa

South Africa has more people with HIV/AIDS than almost any other country in the world (one in every six) but, before 1982, there were none. By 1990, 2% of all adults in the country were infected and, by 2000, this had increased to 20%. In 2010 the infection rate had dipped to 18% but 800 people were still dying each day from the disease.

Although drugs are available to treat AIDS, in South Africa there are **not enough trained staff to administer the treatments** and there are **many very isolated areas** which are difficult to reach. Many people do not know they have the disease. **Testing facilities are poor** and many **people avoid testing** because of the stigma associated with the disease.

In 2000, the government began a big recruitment and training programme for medical staff, and ARV drugs were made available much more cheaply (£100 per patient per year). In addition, 160 million free condoms were distributed.

Since 1998, there have been **HIV education campaigns**, informing people about AIDS and the need for safe sex. This has been made difficult because **one in seven South Africans cannot read** and there are **eleven official languages** in the country. To overcome this, the campaign uses radio, TV soap operas and drama. None of South Africa's strategies for dealing with AIDS began until the disease was already rife in the country and this has made it much more difficult to control.

Did you know...?
A study in 2004 found that South Africans spend more time attending funerals than shopping.

National 4

1. What are the three main ways in which people contract HIV/AIDS?
2. Why is AIDS more common in developing countries than in developed countries?
3. Why does AIDS spread more quickly in countries at war?
4. Choose one of the consequences of HIV/AIDS and explain it.
5. Describe the ways in which AIDS can be treated.
6. List four problems in trying to treat AIDS in South Africa.

National 5

1. Explain fully why AIDS is more common in developing countries than in developed countries.
2. AIDS spreads more quickly in countries at war. Explain why.
3. In a country in which many people suffer from HIV/AIDS, explain why:
   (a) the children are less educated
   (b) there is less tax money
   (c) there is less wealth produced.
4. Describe in detail the ways in which AIDS can be treated.
5. Describe three problems in treating AIDS in South Africa.

Activities

Activity A

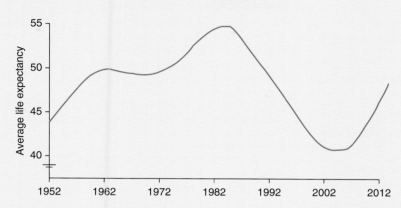

The graph above shows changes in life expectancy in a typical country in southern Africa over the last 50 years.

(a) Draw the graph.
(b) Decide when the following events took place in the country shown and write them in the correct place on the graph:

- ARV drugs available
- Free primary education introduced
- First case of HIV/AIDS
- Short civil war
- Free condoms available

- New water supply schemes
- Trade deal with the EU
- Drought
- Poverty increasing rapidly
- MP says garlic cures AIDS

Activity B

Below is a completed crossword but with the clues missing. You must come up with a clue for each of the words in the crossword to describe what each word means in relation to HIV/AIDS.

							¹B			²P	
	³M	⁴A	L	N	U	T	R	I	T	I	O	N
		I					E			V	
⁵N	E	E	D	L	E		A			E	
		S					S			R	
			⁶T	R	E	A	T	M	E	N	T
							M			Y	
	⁷I	N	F	E	C	T	I	O	N	S	
							L				
							K				

Now complete the 'I can do' boxes for this chapter.

'I can do' checklist

	Red	Yellow	Green	Comment
Chapter 1 Climate change				
Describe what climate change is				
Describe the natural greenhouse effect				
Give examples of evidence that our climate has changed				
Chapter 2 Climate change – physical factors				
Describe how physical factors affect the world's climate				
Give examples of some physical factors that cause climate change				
Explain how these factors cause climate change				
Chapter 3 Climate change – human factors				
Describe the difference between the natural and the enhanced greenhouse effect				
Give examples of some human activities that contribute to climate change				
Explain how these activities affect the climate				
Chapter 4 Climate change – its effects				
Give examples of some of the ways that changes in the climate can positively affect countries				
Give examples of some of the ways that changes in the climate can negatively affect countries				
Chapter 5 Climate change – coping with its effects				
Give examples of strategies to reduce the effects of climate change at a local level				
Describe some strategies that can be used on a national level to reduce the effects of climate change				
Give examples of strategies to reduce the effects of climate change on an international level				
Chapter 6 Climate change – case study of Bangladesh				
Describe the human and physical geography of Bangladesh				
Explain some effects of climate change on Bangladesh				
Give reasons why Bangladesh finds it difficult to deal with the impact of climate change				

Chapter 7 Climate change – case study of Florida, USA				
Describe the human and physical geography of Florida				
Explain some of the effects of climate change on Florida				
Give reasons why Florida finds it easier to deal with the impacts of climate change				
Chapter 8 Structure of the Earth				
Explain the meaning of the term *natural (environmental) hazard*				
Describe the three main layers of the Earth				
Describe the two types of the Earth's crust				
Chapter 9 Crustal plates and plate boundaries				
Give a definition of a crustal plate				
Describe the four types of plate boundary				
Describe the activities which take place at each plate boundary				
Chapter 10 Volcanoes				
Describe the location of volcanoes around the world				
Explain the formation of volcanoes at plate boundaries				
Describe the features of a volcano				
Chapter 11 The eruption of Mt. St. Helens, 1980				
Explain the reason for the eruption				
Give examples of the effects of the eruption on the landscape				
Describe the impact of the eruption on the people				
Chapter 12 Managing the eruption of Mt. St. Helens, 1980				
Give examples of help that was given before the eruption				
Describe some of the short-term aid needed following the eruption				
Describe what long-term aid is and why it was needed				
Chapter 13 Earthquakes				
Describe where earthquakes occur				
Explain how an earthquake happens				
Describe how earthquakes are measured				

Chapter 14 The cause of the Japan earthquake, 2011				
Explain why Japan experiences earthquakes				
Describe the cause of the 2011 earthquake				
Give reasons why predicting earthquakes is difficult				
Chapter 15 The effects and management of the Japan earthquake, 2011				
Describe the impact of the earthquake on the landscape				
Give examples of how the earthquake affected the people of Japan				
Describe how Japan coped with this earthquake				
Chapter 16 Tropical storms				
Describe what a tropical storm is				
Describe the main features and locations of tropical storms				
Describe the conditions needed to create a tropical storm				
Chapter 17 The cause of Hurricane Katrina, 2005				
Describe the differences between tropical storms and hurricanes				
Explain the conditions that formed Hurricane Katrina				
Describe the path of Hurricane Katrina				
Chapter 18 The impact and management of Hurricane Katrina, 2005				
Explain the impact of Hurricane Katrina on the landscape				
Give examples of the impact of Hurricane Katrina on the people				
Describe the relief effort after Hurricane Katrina				
Chapter 19 Global differences in employment				
Give examples of primary, secondary and tertiary industry				
Describe how the types of industry in a country change over time				
Describe the countries which are important for the three types of industry				
Chapter 20 World trade patterns				
Explain the differences in the imports and exports of developed countries and developing countries				
Describe the main barriers to world trade				
Explain how world trade is unfair to developing countries				

Chapter 21 The growth in world trade				
Explain why countries trade more now				
Understand the meaning of countries being *interdependent*				
Understand the meaning of *globalisation*				
Chapter 22 Globalisation				
Describe how globalisation benefits people in rich and poor countries				
Describe the problems globalisation brings to people and countries				
Explain the effects of globalisation on the environment				
Chapter 23 Transnational corporations				
Give a definition and examples of transnational corporations (TNCs)				
Describe the benefits TNCs bring to countries				
Describe the main criticisms of TNCs				
Chapter 24 Coca-Cola – a transnational corporation				
Give evidence that Coca-Cola is a TNC				
Describe the benefits that Coca-Cola, as a TNC, has brought to other countries				
Describe some of the problems Coca-Cola has created in other countries				
Chapter 25 Making international trade fair				
Describe the advantages and disadvantages of trade alliances				
Give a definition of *sustainable trade practices*				
Describe one or more examples of sustainable trade practice				
Chapter 26 South Korea – a rapidly changing economy				
Describe how employment in South Korea has changed since 1950				
Give reasons why South Korea has been able to industrialise				
Give examples of good and bad effects of industrialisation on South Korea				
Chapter 27 Health in developing countries				
Describe the main reasons for ill-health in developing countries				
Explain the effects of ill-health on the people and their countries				
Describe some solutions to ill-health in developing countries				

Chapter 28 Health in developed countries				
Explain why developed countries have less disease than developing countries				
Describe the areas within developed countries which have the worst health				
Describe the main factors affecting health in developed countries				
Chapter 29 Malaria – its cause and transmission				
Describe the distribution of malaria around the world				
Explain the causes of malaria				
Describe how malaria is transmitted				
Chapter 30 Malaria – factors in its distribution				
Describe the physical environment where malaria is likely to be found				
Explain the human factors that contribute to the spread of malaria				
Give examples of methods used to control malaria				
Chapter 31 Heart disease – its causes				
Describe the main causes of heart disease				
Draw and interpret graphical data				
Chapter 32 Heart disease – methods of control				
Describe the factors in the distribution of heart disease				
Give examples of methods of controlling heart disease				
Explain the role played by the National Health Service in preventing and treating heart disease				
Chapter 33 HIV/AIDS – its distribution				
Explain how HIV/AIDS affects people				
Describe the global distribution of HIV/AIDS				
Draw and interpret choropleth maps				
Chapter 34 HIV/AIDS – causes, effects and treatment				
Describe the factors affecting the distribution of HIV/AIDS				
Explain the consequences of HIV/AIDS for a country				
Give examples of how HIV/AIDS can be treated				

Index